KT-442-108

B.S.U.C. - LIBRARY

00303978

GENTLEMEN CALLERS

GENTLEMEN CALLERS

Tennessee Williams, Homosexuality, and Mid-Twentieth-Century Broadway Drama

Michael Paller

GENTLEMEN CALLERS
Copyright © Michael Paller 2005

All rights reserved. No part of this book may be used or reproduced in
any manner whatsoever without written permission except in the case of
brief quotations embodied in critical articles or reviews.

First published 2005 by
PALGRAVE MACMILLAN™
175 Fifth Avenue, New York, N.Y. 10010 and
Houndmills, Basingstoke, Hampshire, England RG21 6XS.
Companies and representatives throughout the world.

PALGRAVE MACMILLAN is the global academic imprint of the
Palgrave Macmillan division of St. Martin's Press, LLC and of Palgrave
Macmillan Ltd. Macmillan® is a registered trademark in the United
States, United Kingdom and other countries. Palgrave is a registered
trademark in the European Union and other countries.

ISBN 1-4039-6775-X hardback

Library of Congress Cataloging-in-Publication Data
Paller, Michael.
    Gentlemen callers : Tennessee Williams, homosexuality, and mid-
twentieth-century drama / Michael Paller.
        p.   cm.
    Includes bibliographical references and index.
    ISBN 1-4039-6775-D
    1. Williams, Tennessee, 1911–1983—Criticism and interpretation.
2. Homosexuality and literature—United States—History—20th
century.   3. Homosexuality, Male, in literature.   4. Sexual orientation
in literature.   5. Gay men in literature.   6. Sex in literature.   I.
Title.
PS3545.I5365Z799   2005
812'.54—dc22
                      2004054129

A catalogue record for this book is available from the British Library.

Design by Letra Libre.

First edition: April 2005

10 9 8 7 6 5 4 3 2 1
Printed in the United States of America

Transferred to digital printing in 2006

BATH SPA UNIVERSITY
NEWTON PARK LIBRARY
Class No.

Contents

In memory of Bradley Ball
1960–1995

and
for Steven

Acknowledgments

My first thanks are to the historians who have been reclaiming gay and lesbian history for the last three decades. Anyone writing about gays and lesbians in the years before 1969 owes them an incalculable debt. The works of Jonathan Ned Katz are monuments of research; so, too, are the books of John D'Emilio, to which I turned constantly. Kaier Curtin's work on gay and lesbian plays produced on Broadway contains valuable information not found elsewhere.

Second is the community of Tennessee Williams scholars who have been so gracious and helpful: Allean Hale, Janet Haedicke, and Ralph Voss all read a late version of the manuscript and their encouragement and advice helped me finally get over the hump. Robert Bray, editor of *The Tennessee Williams Annual Review,* kindly gave permission to reprint material that now appears in different form in chapter 4. Nicholas Moschovakis shared generously his knowledge on unpublished one-acts; Colby Kullman, Jack Barbera, Barbara Harris, and Nancy Tischler also freely lent their expertise. Kenneth Holditch, who knows more about Southern literature than anyone, probably, shared his many insights into Williams gleaned not just through scholarship but personal friendship with the playwright; he also provided personal tours of some of New Orleans' finest restaurants—no small thing for a writer from New York. At New Directions, Peggy Fox and Thomas Keith long have been invaluable in steering me through the thickets of published and unpublished Williamsiana.

At the Rare Book and Manuscript Division of Butler Library at Columbia University I owe a debt to Jean Ashton, Bernard Crystal, and their staff; at the UCLA Library Department of Special Collections, Charlotte Brown and Genie Geurard; and the staff of the New York Public Library for the Performing Arts, one of America's great cultural treasures. Patricia Hepplewhite generously granted permission to quote letters from her brother, the late David Lobdell, about his friend Tennessee Williams. Sarah Schulman and Michael Schiavi also read portions of the manuscript and offered much-needed criticism and research advice.

John Crowley was my teacher at Syracuse University; for more than 20 years he has been my friend. He urged me on from the very beginning of this project and encouraged me when I was bogged down and discouraged, saw me through some very difficult personal crises and always offered the best kind of professional and personal advice: honest and constructive.

I discovered Eric Bentley's books just after graduating from college and read them as quickly as I could buy them. I aspired to the clarity of their expression and the audaciousness with which they stood conventional wisdom on its head. I still do. So it is with much gratitude that I thank him for his friendship and for the shrewd comments he made on the chapters of the manuscript that he read, and also for all of those Saturday night dinners.

I thank my editor at Palgrave Macmillan, Farideh Koohi-Kamali, for having faith in this book and for making such a passionate case for it to her colleagues.

When I started writing this book, my sister, Francie Newfield, gave me a set of *The Theatre of Tennessee Williams* as a fortieth birthday present. A decade later, I have something to show for it. My first awareness of theatre, it seems, came when I hurled a block at her head because she and not her three-year-old brother was accompanying our parents to the road company of *My Fair Lady* when it visited Cleveland in 1958. However, Betty Rose and Orrie Paller made up for this lapse hundreds of times over by feeding our appetites for all the arts, and especially for reading. What greater gift can parents give their children than new ways of experiencing and thinking about the world?

Finally, two men. Bradley Ball was there when this book had its first incarnation as my Master's Thesis. He died of complications from AIDS in January 1995 at the age of 34, shortly after my graduation from Columbia and his own from Fordham. Theatre, politics, and gay rights activism were among his passions and some of him, I hope, is in this book. In 1999 I met Steven Melvin. The stability of his love (not to mention his editorial expertise) gave me the strength to finish what I started—but that's just the beginning of what he means to me.

New York City
September 2004

The author would like to thank the following for permission to quote from published and unpublished material: "Auto-Da-Fé," "Lord Byron's Love Letter," "Something Unspoken," by Tennessee Williams, from *27 Wagons Full of Cotton*, copyright ©1945 by The University of the South. Reprinted by permission of New Directions Publishing Corp and Georges Borchardt, Inc., for the author. *Camino Real* by Tennessee Williams, from *Camino Real*, copyright © as "Camino Real," revised and published version, ©1953 by The University of the South. Renewed 1981 The University of the South. Reprinted by permission of New Directions Publishing Corp and Georges Borchardt, Inc., for the author. *Cat on a Hot Tin Roof* by Tennessee Williams, from *Cat on a Hot Tin Roof*, copyright ©1954, 1955, 1971, 1975 by The University of the South. Reprinted by permission of New Directions Publishing Corp and Georges Borchardt, Inc., for the author. "Mornings on Bourbon Street" by Tennessee Williams, from *The Collected Poems of Tennessee Williams*, copyright ©1977 by The University of the South. Reprinted by permission of New Directions Publishing Corp and Georges Borchardt, Inc., for the author. "Three Players of a Summer Game" by Tennessee Williams, from *The Collected Stories of Tennessee Williams*, copyright ©1948 by The University of the South. Reprinted by permission of New Directions Publishing Corp and Georges Borchardt, Inc., for the author. *The Glass Menagerie* by Tennessee Williams, from *The Glass Menagerie*, copyright ©1945 by The University of the South and Edwin D.

Williams. Reprinted by permission of New Directions Publishing Corp and Georges Borchardt, Inc., for the author. *Something Cloudy, Something* Clear by Tennessee Williams, from *Something Cloudy, Something Clear,* copyright ©1995 by The University of the South. Reprinted by permission of New Directions Publishing Corp and Georges Borchardt, Inc., for the author. *Suddenly Last Summer* by Tennessee Williams, from *Suddenly Last Summer,* copyright ©1958 by The University of the South. Reprinted by permission of New Directions Publishing Corp and Georges Borchardt, Inc., for the author. *Small Craft Warnings* by Tennessee Williams, from *The Theatre of Tennessee Williams Vol. V,* copyright ©1963, 1964 by Two Rivers Enterprises, Inc. Reprinted by permission of New Directions Publishing Corp and Georges Borchardt, Inc., for the author. *Vieux Carr* by Tennessee Williams, from *The Theatre of Tennessee Williams Vol. VIII,* copyright ©1977, 1979 by The University of the South. Reprinted by permission of New Directions Publishing Corp. and Georges Borchardt, Inc., for the author. Unpublished letters of Tennessee Williams and extracts from *Memoirs,* copyright ©2004 by The University of the South. Reprinted by permission of Georges Borchardt, Inc., for the Tennessee Williams Estate. For material in chapter 4 regarding Williams's psychoanalysis, *The Tennessee Williams Annual Review* Number 3 where it appeared in different form; Beacon Press, Boston, for material from *Radically Gay* by Will Roscoe, copyright ©1996 by Harry Hay and Will Roscoe; for excerpts from "Tennessee Williams, interviewed by George Whitmore," Gay Sunshine Press, copyright ©1991.

Every effort has been made to contact all copyright holders. The author will be happy to make good in future editions any errors or omissions brought to his attention.

Note on quotations: Tennessee Williams was very free and idiosyncratic in his use of ellipses. To distinguish his from mine, the latter appear in brackets.

GENTLEMEN CALLERS

Introduction

In 1949, when the esteemed critic and director Harold Clurman approached several publishers with the idea of writing a biography of Eugene O'Neill, one dismissed him with, "Who cares about him today?" Had it not been for José Quintero's productions of *The Iceman Cometh* and *Long Day's Journey Into Night* in 1956, we might still not care about him. It is a truism that an American writer's reputation often reaches its nadir near or at the time of his death. Herman Melville is only the most extreme example; F. Scott Fitzgerald and Ernest Hemingway were also considered old-hat when they died, if they were considered at all.[1]

Tennessee Williams suffered a similar fate, all the worse because critics didn't wait until he was dead to demand the embalming fluid. The period of his nadir was gratuitously long and cruel. It began when he had 20 years to live, in 1963, with the failure of *The Milk Train Doesn't Stop Here Anymore*. The play opened during a newspaper strike and managed to scratch out 69 performances. It heralded an aesthetic departure and the embrace of new themes, and Williams, determined to improve it, worked hard on a revision. The new version opened on Broadway early the next year, but had the ill luck to appear while the critics were working. It lasted three nights. Was its failure the critics' fault? No; *Milk Train* is a flawed play in either version. But it is a critic's task to sense significant changes in the direction of a major playwright's work and alert the audience that something new is going on, whether he thinks it succeeds or not. Since most theatre critics don't look beneath a play's surface, however, those in 1964 largely assumed that Williams had lost his grip, and said so to their readers, who stayed away.

From then until his death, Williams endured one commercial and critical failure after another. More than that, something about Williams's post–*Milk Train* work brought out a viciousness just barely latent in many critics, both journalistic and academic. The critics' confusion was avoidable but understandable, if their vitriol wasn't. Each new play that was not *The Glass Menagerie* or *A Streetcar Named Desire*—each one that failed to earn a profit or instantly reveal its meaning—brought a chorus of bitter attacks, as if critics regarded each new work as a personal affront. (See, for example, in chapter 6, Stefan Kanfer's review of *In the Bar of a Tokyo Hotel*. To him falls the honor of writing what may be the most vindictive review ever perpetrated on a major American playwright.)

What accounts for the malice with which critics treated Williams, who, if he had written nothing else, had given the world *The Glass Menagerie, Streetcar,* and *Cat on a Hot Tin Roof?* I contend in chapters 5 and 6 that homophobia had more than a little to do with it. It was never a secret that Williams was gay, and from *Streetcar* on, some critics used this knowledge against him. They also used it against those characters who were gay, by either trivializing their importance or ignoring them altogether. One should not be surprised that straight critics writing between the 1940s and 1970s would behave this way, sometimes out of malice, more often from a simple lack of understanding or even vocabulary. What is stranger is that, in the 1990s, some gay critics took up where their straight predecessors left off. Since then, the problem has been that Tennessee Williams isn't gay *enough;* that he was incapable of producing a "positive image" of a gay person. One of the purposes of this book (also by a gay critic) is to show that none of Williams's gay characters is simply the product of its author's alleged self-loathing, but an amalgam of personal, social, and historical forces. Indeed, I hope to show that any self-loathing that Williams may have had was minimal, and that his work was the place where he struggled with and overcame it.

Evolutions in culture frequently change what a work seems to mean, as if the language it was originally written in has been lost. Sometimes these changes help us understand what we might have missed the first time. Michael Kahn's 1974 revival of *Cat on a Hot Tin Roof* at the American Shakespeare Festival, for example, made clear to Walter Kerr of *The New York Times* what he thought had been evasions on Williams's part in 1955. "Was [Brick] or was he not homosexual, did Williams mean him to be but—given the discreet silences of twenty years ago—cautiously refuse to say so?" The answer, Kerr now decided, was yes.[2]

Often, however, time—which Williams declared to be his enemy from the very beginning—takes us further away from a playwright's intentions. Williams's plays with gay characters come to us from a world that is quite different from our own, and were it not for the efforts of gay and lesbian historians to recover that era, those of us born after 1950 would know nothing of it at all. Thirty years of gay liberation, the rise of identity politics, and the creeping assertion in academic and theatrical circles that judgments based on a work's supposed political utility are the most, if not the only, legitimate judgments, have interfered with our ability to see these plays as Williams meant them to be seen. So when we demand that plays written in the 1940s and 1950s fit a contemporary notion of what image a gay character ought to project, we're unlikely to arrive at an assessment that's fair or honest—especially when that character is considered apart from any other aspect of the play. To judge these plays accurately, we should know what Williams intended to say about gay life—if anything—and something about the world he was saying it to. This calls for a careful study of the texts balanced by a sufficiently detailed examination of the cultural, personal, and political times and circumstances of their creation. This book, then, examines the gay characters and, in one case, the lesbian characters, in Williams's plays through the compound lens of his life and times. As such, it is as much a cultural history as a work of criticism.

I don't presume to think that these plays need rescuing by me. My intention is to help readers, theatregoers, actors, and directors recall what Williams intended them to say. Then, we'll be in a better position to judge how they speak to us at our moment in time.

Due to space limitations, I cannot consider *A Streetcar Named Desire* here. Some would say that's fine, as there are no gay characters in it, anyway. But Allan Grey is a major figure in that play, although he is offstage and dead—a fact used by some scholars to claim the play is an exercise in self-loathing and homophobia. I would like to have mentioned, too, the treatment of gay and lesbian members of the armed forces during and after World War II; it is an important part of the context in which one must consider the Allan Greys of America, circa 1940. I have written about the play in *The Tennessee Williams Literary Journal,* Spring 2003, and for now those who are interested can read about it there. For the story of gay men and lesbians during World War II, there is the excellent *Coming Out Under Fire* by Allan Bérubé.

One more thing: If nothing else, the publisher's response to Clurman's offer to write a biography of O'Neill suggests that a playwright's worth cannot be proven or disproven definitively in a publisher's office, or in the pages of a newspaper, scholarly journal, or book. Only in the theatre can his or her work accurately be measured—and even then only when the production and the audience are equal to the play.

ONE

The Signs Are Interior

I

In September 1941, Tennessee Williams returned to New Orleans. "The second New Orleans period here commences," he wrote in his journal. Nineteen months had passed since his first, abortive, visit: That period had lasted a scant two months, January and February, in 1939. In the intervening months, Williams had experienced the exhilaration of seeing *Battle of Angels* optioned and staged by the Theatre Guild, the "prestige" theatre of Broadway, and suffered the humiliating nightmare of its frigid reception in Boston. The few audience members who remained when the curtain came down after the first performance sat in hostile silence. A week later, *Battle of Angels* closed. It did not come to New York.[1]

Also in those months, Williams finally recognized the nature of his sexuality and had what appears to have been his first homosexual experience. How reconciled he would become to his homosexuality is a matter of dispute; there is evidence on both sides. What one can say with certainty is that this acknowledgment was arrived at only after a years-long struggle. Williams was now 29 and, in all likelihood, he'd had one physically consummated relationship with a woman, a fellow theatre student at the University of Iowa. His relationships with men weren't many more.

The event precipitating the end of Williams's first New Orleans sojourn was Mardi Gras. Traditionally during that week, inhibitions—not strong to begin with in the Crescent City—went entirely by the boards. The playwright loathed the bacchanalian atmosphere. Lyle Leverich, Williams's most thorough biographer, speculates that the atmosphere in New Orleans generally, and the frenzy of Mardis Gras in particular, edged Williams closer to the recognition of his homosexuality—a recognition for which he was not entirely prepared. And so Williams did what he would habitually do when feeling psychologically threatened: He fled.

Still, the city made an impression on him. He said later that in New Orleans, "I found the kind of freedom I had always needed. And the shock of it against the Puritanism of my nature has given me a theme, which I have never ceased exploiting." The tension created by this collision of freedom and the consequent shock, the instinctive lure of sex and an almost ingrained repulsion by it, would be necessary before Williams could do his best work.[2]

Williams's adjustment to his homosexuality was not made easier by a love affair that had left him deeply scarred. In Provincetown during the summer of 1940, the young playwright fell in love with a Canadian dancer, Kip Kiernan, who for a time returned Williams's affections. Barely a month had passed, however, when Kip told the love-starved, sexually aggressive playwright that the affair was over. The dancer had been warned by a woman friend that he was in danger of being turned into a homosexual, and he didn't want that. He broke with Williams and, in a rage, Williams hurled a boot at the girl who had destroyed his happiness. The intensity of his desire and the nearly insatiable sexual appetite it awoke was as shocking to him as Kip's rejection was devastating. From this moment, Williams would protect his heart against such an intensity of longing and hurt, yet he would do so with regret. In *Vieux Carré*, written in the last decade of Williams's life, The Writer, a young man coming to terms with his homosexuality and need for love, would say, "You have to protect your heart." Nightingale, an older painter bent on seduction, remonstrates, "With a shell of calcium? Would that improve your work?"[3]

———

The circumstances of Williams's childhood made his sexual formation more fraught with anxiety than many of the other men and women born in the deep South in the early years of the last century. His mother, Edwina Dakin Williams, was the deeply repressed, puritanical, often hysterical daughter of an Episcopal minister. Williams's father, Cornelius Coffin Williams, came from pioneering Tennessee stock and temperamentally was Edwina's exact opposite, as unlikely a husband for her as could be imagined. Cornelius's mother died when he was five, leaving him, as Williams wrote, without the "emolient influence of a mother." It became nearly impossible for Cornelius to express love or other tender emotions. He grew into a hard-drinking, profane, restless young man who served in the Spanish-American War and, at the time of his marriage to Edwina, was working as a manager for the Cumberland Telephone and Telegraph Company. Very soon, like his absent stand-in in *The Glass Menagerie,* he fell in love with long distance, and traded his desk job for the itinerant life of a traveling salesman. C. C., as he called himself, was happiest on the road, showing his wares—men's clothing—and playing cards and drinking with his fellow drummers. His stays at home with Edwina, in her father's rectory in Columbus, Mississippi, were stuffy and confining, and they grew shorter and shorter. He came to resent the Episcopal propriety and uprightness of rectory life. Worse, Edwina hated sex and submitted to it (when she did) with the most reluctant sense of duty. In time, sex between them would stop altogether and they would keep separate bedrooms. In later years, Williams remembered that as a boy he would hear terrible sounds coming from their room as Cornelius forced himself on his wife. Eventually, the playwright would recognize both sides of the family in himself: the Dakin sensitivity and puritanism, and the Williams aggression and toughness.[4]

During his earliest years, however, it was Edwina who set the tone in the Williams house, and it was stifling. She succeeded in passing on to all

three children her own guilt and shame regarding sex. Her fanatical hatred of sex would contribute heavily to the mental illness of her daughter
Rose; her younger son, Dakin, would be a virgin when he married in his
late thirties.

It is hardly a surprise, then, that Tennessee Williams would struggle
for years over the issue of his sexual identity. Until he was past his middle
twenties he was attracted to women as well as to men. Still, lonely and
alienated as he was, sex with a woman could be deeply disturbing, and
his years at college did not clarify matters. Indeed, they only seemed to
confuse him further. In *Tom,* Lyle Leverich writes that while Williams
began to experience a growing attraction to his own sex at college, he
could not understand it himself. He tried to keep his increasingly homoerotic impulses at bay by surrounding himself with girls.[5]

Edwina Williams's neurotic attitudes imposed on Tom a deep ambivalence about sex. This feeling, it should be noted, was not about homosexuality; it was about *sexuality.* It was a tangled nexus of feelings from
which he never entirely freed himself—and it has led some critics to mis-
takenly conclude that he was filled with self-loathing.

In June 1939, four months after leaving New Orleans the first time,
Williams had the first homosexual experience he mentions in any of his
journals. It was traumatic. He wrote that it "confused and upset me and
left me with a feeling of spiritual nausea." For days the memory of it
haunted him and plagued him with guilt. Very soon, however, gay sex
would nauseate Williams no longer. By the end of July he was infatuated
with Kip. Writing to his friend Donald Windham in late June 1940,
Williams describes sex with Kip in ecstatic terms. He compares his body
to Greek statuary and the Statue of Liberty and details the intimacies
they shared two or three times a night. As important as the physical sensation, however, was Williams's conviction that Kip loved him, and when
this proved to be an illusion, the consequences were disastrous.[6]

According to Windham, with whom the playwright would soon be
cruising Times Square, and who knew the playwright better than anyone
else in those years,

> Tennessee was full of guilt about sex; he was full of guilt about numerous
> aspects of his life and character; but homosexuality, being homosexual, en-

joying the friendship of people who shared his desires, was what he loved more than he loved any specific persons, more than he loved anything, in fact, except his writing.[7]

But the emotional intensity that would accompany the break-up with Kip left him doubting his sexual identity again. He wrote in a journal entry that he expected to wind up with a woman. "I feel now it will be a woman I will finally go to for tenderness in life. The sexual part—if there has to be any—would probably adjust itself in a while, since I am so easily directed in that way." Even here, he is reluctant to embrace any kind of sexuality: *if there has to be any.*[8]

The roots of sexuality, as well as one's response to it, are complicated affairs in any individual. They are made more complicated by the society in which one lives and from which one receives hundreds of daily messages about what constitutes masculine and feminine, proper and improper, behavior. Williams was born into a sexually dysfunctional family amid a deeply conservative Southern society in 1911, and to argue that this environment did him no psychological harm, or that Williams lacked any trace of homophobia, would be foolish. The society that provided Tennessee Williams with so much of his material also did him considerable psychic damage. More important than the homophobia Williams could not help but absorb, however, was his battle against the myriad examples that would confront him. Perhaps he never entirely accepted his homosexuality; but he also rejected the prejudice against it that was so pervasive in his lifetime as to be practically invisible.

If one wanted to formulate the intricate, inevitable associations between Williams, the society he lived in, and his work, the formula would be this: The society into which Williams was born was deeply homophobic. Williams could not help but absorb some of those homophobic attitudes. But he did not accept them. Rather, he fought against them and his struggle is reflected in his work. A corollary to that formula then suggests itself: Williams's struggle against the homophobia that surrounded him transmuted that homophobia into something subtler and more complex. Homophobia and self-loathing became part of a whole continuum of feeling regarding his sexuality, from revulsion to joy, from negative to positive, but in no orderly, linear progression. This continuum was messy, dynamic, and unruly; it would turn back on itself, and move ahead again in the direction we would consider "progress" (that is, from less accepting to more) before backtracking, and then making "progress" once more. Perhaps, indeed, it was less a continuum than an evolutionary ebb and flow.

This battle for self-acceptance is one that Williams probably would have been happy to avoid, could he have done so. But the struggle occurred, and the creative work that resulted was deep and rich. The struggle encompassed a world as broad as the society in which he lived and as intimate as the acts of creation with which he responded.

The curious thing is that to read his *Memoirs* and other autobiographical writings is to get the impression that no such battle ever was fought. The book leaves the impression that Williams recognized his homosexuality even as a teenager and, although he was also attracted to girls, he accepted his condition. By the time he went to college, he writes, he had accepted his gayness. He mentions appreciating the bodies of boys when he was 15. He relates an episode at college in which a sleepwalking roommate stumbled into the young writer's bed, cried out in dismay, then stumbled away. Williams concludes this story by saying that he waited for several nights, hoping that the incident would happen again. Then there was the infatuation with a college roommate he described as having "very large and luminous green eyes," although he writes that the infatuation was "a mostly sublimated attachment [. . . .]" He portrays himself as being so well-adjusted to his homosexuality in those years that he could even give advice to a troubled friend: When a fellow student confessed to him about going to another man's room and wanting to touch him "like a man does a woman," Williams had a sage answer. "When I finally spoke, I said to him, 'There's nothing about it to be upset over.'" The friend responds that his feelings were unnatural. But, Williams tells him, "It's perfectly natural and you are just being silly." Lyle Leverich, combing through Williams's journals and conducting interviews with people who knew him during his college years, could confirm none of these episodes.[9]

In his *Memoirs*, Williams never hesitates to discuss his homosexuality, but the pain and confusion he experienced while coming to understand his nature are completely elided. Perhaps he omitted them because after a lifetime the pain and confusion were still sharp. Perhaps he wanted to be known as the "founding father of the uncloseted gay world," as he called himself in a letter to Maria St. Just, and founding fathers are not supposed to harbor doubts. Perhaps he was unaware that these events had never occurred: He would hardly be the first person to unconsciously rewrite his own history. "I suspect what I am haunted by is something that I am concealing from myself, unconsciously but wisely," he wrote in the *Memoirs*. Whatever the reason for speaking of his early awareness of his homosexuality, Williams is engaging in what had been, from his earliest work, his principal dramatic strategy: to reveal a little while concealing a great deal more. There is a fundamental tension, in other words, found in Williams's best plays, between the need to reveal and the urge to conceal. However, far from being the serious flaw that some critics interpret it to be, this tension proved to be not only necessary, but fruitful and positive.[10]

II

When Williams returned to New Orleans in 1941, opportunities abounded for a young gay man to satisfy his sexual appetite while protecting his heart. On Labor Day, September 1, President Roosevelt declared in a radio broadcast that, "we must do our full part" to conquer Hitler's "forces of insane violence." "There has never been a moment in our history," he declared, "when Americans were not ready to stand up as free men and fight for their rights." New Orleans was ready to do its part. Three hundred thousand National Guard troops were billeted upstate in Alexandria and regularly traveled to New Orleans on weekend passes, increasing the passenger traffic on the railroads entering the city 20 percent over the previous year. The shipyards along the industrial canal were expanded and rebuilt, providing jobs for 1,650 men; in November, some 33 Liberty ships were scheduled to be built. In addition, on the average, 23 passenger and freight ships were arriving in the port each week. The New Orleans autumn of 1941 offered Tennessee Williams many opportunities for temptation, and Williams, whatever his ambivalence, rarely had trouble giving in. However, the city was also awash in military police and the shore patrol, whose job it was to keep their men out of trouble, and out of gay gathering places. Gay bars were frequently raided, and Williams often barely escaped arrest.[11]

Despite the sexual distractions New Orleans offered and the sharp mixture of pleasure and guilt they provided, Williams devoted most of his days to writing. He was slowly acquiring a name for himself, but he was still quite poor. For a few weeks, he subsisted on checks from the actor Hume Cronyn, who had optioned some of his one-acts. The checks were only intermittent; more than once he had to hock his typewriter and write with a pencil. He moved from rooming house to rooming house, barely scraping together the rent. One such rooming house was on Royal Street. Williams wrote to his friend Paul Bigelow, "'Cher,' I have a room on Royal right opposite *the* gay bar—The St. James, so I can hover *like* a bright angel over the troubled waters of homosociety and I have a balcony and everything but a mantilla to throw across it. But I *do* wish you would mail me my laundry. You don't want people to whisper—'The poor girl's putting up a very bold front, but actually doesn't have a *shirt* to her *back!*" Bigelow mailed the laundry that Williams had left behind during a stay in New York. Indeed, he mailed it several times to each of Williams's new addresses. Each time, it was returned to New York unclaimed.[12]

Williams's attitude toward sex and homosexuality varied almost as much as his address. He was continually attracted to gay bars like the St. James, and almost as often, came away filled with loathing. The struggle occupied much of the time he did not devote to writing, which he did steadfastly, every day. His journal reflects the intensity of the battle between his puritan nature and his powerful desires. In September, he records a discussion with his friend Oliver Evans, with whom he'd often cruise. After a visit to the St. James, Evans declared, "We ought to be exterminated for the good of society." Williams objected: If homosexuals were exterminated, society would lose most of its artists and its spiritual values. "We are the rotten apples in the barrel," Evans insisted. "We ought to be exterminated at the age of 25." "How many of us feel this way, I wonder?" Williams asked himself:

> Bear this intolerable burden of guilt? To feel some humiliation and a great deal of sorrow at times is inevitable. But feeling guilty is foolish. I am a

deeper and warmer and kinder man for my deviation [. . .] Someday society will take perhaps the suitable action—but I do not believe that it will or should be extermination.—Oh, well.[13]

He sought out sex, usually with success, but it left him unsatisfied. "I had another friend last night," he wrote a short time after his colloquy with Evans. "The cold and beautiful bodies of the young! They spread themselves out like a banquet table, you dine voraciously and afterwards it is like you had eaten nothing but air." Early in October, *"Saturday* Night, I cruised with 3 flaming belles for a while on Canal Street and around the Quarter. They bored and disgusted me so I quit and left Saturday Night to its own vulgar, noisy devices and went upstairs to my big wide comfortable bed [. . . .]"[14] A week later:

> Love-life resumed with a vengeance last night—2 in the night, 1 in the morning. Enjoyed it, the first couple. Then a bit sordid. Ah well, I guess it comes under the heading of fair entertainment. [. . .]
> Love is what makes it still seem nice after the orgasm. Then is when sex becomes art—after the orgasm. One must be an artist to keep it from falling to pieces uglily. Up to then it is simply craftsmanship and of a pretty crude and simple kind. It is also art, of course, when you first meet the person—selecting the attitude and sticking to it.[15]

Sex was a game, a role, a play, in which Williams portrayed a fictional character, reserving the truth of himself for himself. Another week passes. "Had a pretty satisfactory 'roll in the hay' this evening—then a long, dull round of the gay places to kill time. I have nothing to say to these people after I've been to bed with one of them—then it all seems utterly vacuous."[16]

As Williams's attitude toward gay sex alternated between joy and a kind of objectifying indifference, his work began to be constructed around the poles of revealing and concealing. Two of his new friends, Bill Richards and Eloi Bordelon, moved into a rooming house at 722 Toulouse. This was the house in which Williams had rented a room during his first New Orleans period. Williams would visit them there, and later remembered the name of the landlady (new, since his stay), Mrs. Louise Wire, who would lend her name to the landlady in the story, "The Angel in the Alcove" and the play *Vieux Carré*. When, during the seven months of his second New Orleans sojourn, he wrote the one-act *Lord Byron's Love Letter,* he remembered a story he'd heard as a child from his grandfather about an old woman who claimed to have a love letter from the famous Romantic poet; and he would remember, too, his room in the ramshackle house at 722 Toulouse. That room, in which he wrote compulsively, lacked even a door. Separating him from the hallway was only a tattered curtain. A similar curtain plays a crucial role in the one-act comedy. There are no gay characters in *Lord Byron's Love Letter,* but the play is a significant metaphor for the way in which Williams would write about them. Its action is almost entirely concerned with revealing and concealing.

Lord Byron's Love Letter takes place in the French Quarter during Mardi Gras late in the nineteenth century. Two women, one a fortyish spinster named Ariadne, the other a crone simply called The Old Woman, live a threadbare existence in a rundown apartment. What money they have they earn by displaying a love letter allegedly written by Lord Byron and the journal of the woman, Irénée Marguerite de Poitevent, to whom it was written. That is to say, they earn their meager living as Williams himself did: by trying to sell writing.

Two tourists arrive to view the letter. They are a married couple from the Midwest, typical of the "squares" who pervaded Williams's work: insensitive, obtuse, at best emotionally destructive, at worst physically violent. Here, the Husband (the only name Williams gives him; he calls the woman the Matron) is a drunken lout; the wife an earnest dullard. Given the plays that would come later, one might assume that this comedy's conflict would be prototypical: the insensitive squares versus the sensitive, delicate creatures, with affairs ending badly for the latter. But Williams is not interested in that conflict here; it never occurs. Indeed, the contact between these two camps is perfunctory. The Husband, present at the exhibition under duress, falls asleep until the noise of an approaching parade wakens him, at which point he dashes out. The Matron appears to have no more than a vague high-school knowledge of Byron: since he was a poet, even a love letter written by him is bound to be edifying, and of cultural interest. More concerned with her husband's behavior than with the letter, however, she runs after her mate, neglecting to pay the two women for the privilege of seeing Lord Byron's letter. Her conflict is more with her husband, as Ariadne and the Old Woman's is with each other. While Ariadne and the Old Woman are at least spared the physical harm often visited on the sensitive in a Williams play, they experience the emotional violence of seeing what is most meaningful to them, a cherished memento of a lost love, dismissed uncomprehendingly by representatives of an unfeeling majority who favor vulgar, transitory displays. But this is not the energy on which the play's engine runs.

Upon the tourists' appearance, the Old Woman disappears behind some curtains. Or rather, she incompletely disappears. The stage direction reads, *She withdraws gradually behind the curtains. One of her claw-like hands remains visible, holding a curtain slightly open, so that she can watch the visitors.* From her hiding place, she dictates to Ariadne exactly how much of the journal may be read aloud, and how much distance must be kept between the letter and the tourists—making certain that none of the letter's contents can actually be seen.[17]

For her part, Ariadne tries to assert herself and tell the story as she deems fit. But the Old Woman prevails. To Ariadne's increasing irritation, the Old Woman barks out stage directions and embellishments, dominating Ariadne's long-since memorized recitation, which the young woman performs with what appears to be the interest of a bored tour guide:

Matron *[stiffly]*: [. . .] What was Lord Byron doing in Greece, may I ask?
Old Woman *[proudly]*: *Fighting for freedom!*
Spinster: Yes, Lord Byron went to Greece to join the forces that fought against the infidels.
Woman: He gave his life in defense of the universal cause of freedom!
Matron: What was that, did she say?
Spinster *[repeating automatically]:* He gave his life in defense of the universal cause of freedom.
Matron: Oh, how very interesting!
Old Woman: Also, he swam the Hellespont.
Spinster: Yes.
Old Woman: And burned the body of the poet Shelley who was drowned in a storm on the Mediterranean with a volume of Keats in his pocket!
Matron *[incredulously]*: Pardon?
Spinster *[repeating]*: And burned the body of the poet Shelley who was drowned in a storm on the Mediterranean with a volume of Keats in his pocket. (157–8)

The Old Woman tells Ariadne when she may read aloud from the journal with the exhortation, "But please be careful what you choose to read!" She commands Ariadne to show the visitors the young Irénée's picture; corrects her smallest mistakes, and just as Ariadne is about to begin reading, cautions her, "Be *careful!* Remember where to *stop* at, Ariadne!" Ariadne manages to read an entire paragraph uninterrupted before the Old Woman impatiently interjects, "Skip that part! Skip down to where—." At last, the harried spinster manages a small rebellion: "Yes! *Here!* Do let us manage without any more interruptions!" (159–61).

As Ariadne recites from Irénée's journal, the Old Woman becomes swept up in the words. As one stage direction reads, she interrupts Ariadne "in hushed wonder": "Yes—Lord Byron!" and, "The handsomest man that ever walked the earth!" Her impatience gets the best of her again and she insists, as Ariadne tries to read the journal excerpt, "Skip that, it goes on for pages!" and "Go on, skip that, get on to where she *meets* him!" As Ariadne reads Irénée's account of the fateful meeting, the Old Woman erupts in Molly Bloom-like paroxysms: "Yes! Yes! That's the part!" and, a moment later, hoarsely, "*Yes!*" Ariadne continues for another paragraph, before the Old Woman cuts her off in mid-sentence: "Stop *there!* That will be quite enough!.," as if to go any further would unveil a secret that cannot, 70 years after the event, be revealed (162–4).

It isn't until the play's last line, uttered after the Matron and her Husband have run off, that the play's ostensible secret is revealed. In her rage and frustration at the couple's having fled without paying, Ariadne drops Lord Byron's letter on the floor. At last, the Old Woman, stepping from behind the curtains, and, according to the stage directions, *rigid with anger,* exclaims, "Ariadne, my letter! You've dropped my letter! Your grandfather's letter is lying on the floor!"

Establishing a pattern that he would follow in many later plays, Williams never fully reveals the dramatic secret—that bit of information that, since the nineteenth-century well-made plays of Augustin-Eugène Scribe, has been the cornerstone of most Western drama. In *Lord Byron's Love Letter,* the secret that is ostensibly revealed in the curtain line is that Irénée and the Old Woman are one and the same (although in performance we're likely to have guessed it beforehand). But other secrets are never revealed. We never know for certain what Lord Byron's letter contains. The third party—the Midwestern tourists—are never allowed to see it and verify its contents, for themselves or for us.

The play functions as a metaphor for Williams's inner battle between the need to conceal and the need to reveal, with the younger Ariadne seemingly intent on revealing much more of the sexually indiscreet, emotionally luxuriant material than the Old Woman would countenance. Indeed, the Old Woman herself is a paradigm: the curtain behind which she hides conceals her body only partially, while her voice barks instructions that reveal more about the letter and its contents than presumably she would have the world know. In the same way, the young Tennessee Williams, in his first sojourn to New Orleans, perhaps revealed more of his secret soul than he would have the world know, while hidden, in part, behind the curtained doorway to his room at 722 Toulouse.

Lord Byron's Love Letter reflects the first New Orleans experiences that brought Williams close to the realization of his homosexuality when he lived in the rattletrap rooming house in the Vieux Carré. During those two months—months of great poverty, during which he lived as hand-to-mouth as the two women in the play, trying to make a living through writing—he wrestled not only with the demands of money, but with the increasingly obvious demands of his sexual desire. Even as his gayness inevitably dawned on him, he could not give up the attraction he also felt toward women—in part because conventional morality told him that these were the feelings he must have. During those two months, he wrote a poem called "Mornings on Bourbon Street":

Love. Love. Love.

He knew he would say it. But could he believe it again?
He thought of Irene whose body was offered at night
behind the cathedral, whose outspoken pictures were hung
outdoors, in the public square,
as brutal as knuckles smashed into grinning faces.
He thought of the merchant sailor who wrote of the sea,
haltingly, with a huge power locked in a halting tongue—
lost in a tanker off the Florida coast,
the locked and virginal power burned in oil.[18]

Neither Irene nor the merchant sailor was a mere character of Williams's imagination. Irene was a painter in the primitive style who apparently supplemented her meager income with prostitution; the merchant sailor was, before he went to sea, a writer named Joe Turner who worked on the Federal Writers' Project until it closed down. Both seem to have been objects of Williams's desire, and the struggle he surely felt over that desire is reflected in the poem as well as in the subtext of *Lord Byron's Love Letter.* The huge, growing power of Williams's homosexuality was still locked in his halting tongue. But the halting tongue had devised a strategy by which it could reveal what Williams's conscious mind was still bent on concealing—from Williams himself as well as from the world. Williams fled New Orleans the first time at Mardi Gras because its frenzied revelries threatened to unleash the fullness of his sexual desires. *Lord Byron's Love Letter,* in which two women battle over the revelation of a pair of sexual scandals, is set precisely at this time.[19]

If the two women are at odds regarding how, and how much of the Old Woman's scandalous story should be told, then we need to lay out the position of each. This is more complicated than it first appears. It would be natural to connect the Old Woman with repression: She hides behind the curtain and tries to instruct the visible storyteller what to include and, more important, what to exclude. Ariadne, meanwhile, would seem to side with the forces of expression: Her desire throughout the play is to tell the story of Irénée and Lord Byron in her own way, free of the Old Woman's interference.

The Old Woman's interference, however, provides a crucial element that Ariadne lacks: passion. Although the Old Woman rigidly controls how much of the story is revealed to strangers, her emotions are intensely bound up in the telling. Ariadne has told this story hundreds of times and the Old Woman no doubt has repeated it, if only to herself, thousands more. Yet Irénée experiences the emotions of it as if for the first time. Her claw-like hand clutches the curtain with a vise-like grip as her knees, and then the rest of her, give way in ecstasy to the memory of her Byronic tryst.

For Ariadne's part, the sole way she seems capable of repeating this story, which she does only in hopes of survival, is to distance herself from it as much as she can: the stage direction, *repeating automatically,* as she parrots the Old Woman's declaration, "He gave his life in defense of the universal cause of freedom," is crucial. Her recitation is rote; she repeats, verbatim, the Old Woman's words, but without her feeling. What the Old Woman insists on having, and what Ariadne is intent on omitting, is passion.

While representing the urge to conceal, the Old Woman, then, is also the source of intensely held, authentic feelings that society will not sanction. Perhaps only by hiding herself can she reveal them. Her 70-year-old passion for the infamous poet is as sharp now as it was the day she met him, but she cannot bring herself to hear that passion expressed in words before an audience. As for Ariadne, she can tolerate the story only by suppressing the powerful emotions it arouses (and we may be certain that such emotions are indeed aroused in the fortyish spinster: the stage direction, *flurredly,* as she describes Byron's throat, lips, nostrils, and hair, tells us so). In his early plays, this is the way Williams navigated between his urge to write and his fear, not only of sexual desire, but of the profound emotional upheavals that followed. The problem with those earliest plays, however, as Leverich points out, is that Williams omitted from them—as far as he could—himself. The results would be as dry and unconvincing as Ariadne's account of her grandmother's meeting with Lord Byron on the steps of the Acropolis—or, indeed, as *Lord Byron's Love Letter* itself. Drained of the direct expression of emotions, the play provides only sublimation and indirection: comedy.[20]

As early as Williams's first New Orleans sojourn the playwright himself sensed what was missing from his work. He confided to his journal, after receiving the epochal news that his one-acts collectively called *American Blues* had won an award from the Group Theatre:

> My next play will be simple, direct and terrible—a picture of my own heart. There will be no artifice in it. I will speak the truth as I see it—distort as I see distortion—be wild as I am wild—tender as I am tender—mad as I am mad—passionate as I am passionate. It will be myself without concealment or evasion and with a fearless unashamed frontal assault upon life that will leave no room for trepidation . . . a passionate denial of *sham* and a cry for beauty.[21]

Still, he did not write without concealment or a certain amount of evasion. Nonetheless, he was clearing the way for *Battle of Angels*.

The battle to shield himself from the intensity of his feelings—his desire for love as well for sex—would confront him all his life. He would run all the way to Mexico in an attempt to escape the frightening depth of feelings that his love for Kip Kiernan—and Kip's rejection—would reveal to him. He would try to put a shell of calcium around his heart, but the attempt was problematic. He needed a shell to protect himself against his feelings. But how, if he protected himself against his feelings, could he write?

Lord Byron's Love Letter provides us with an outline and a demonstration of the technique that Williams would employ (although not necessarily consciously) for so long and so well: He would confront his urge to conceal with the equally strong need to reveal. Most of Williams's best work would be created through and because of this tension. When the social conventions that required concealment began to ease in the 1970s and the outward necessities that created his conceal/reveal strategy began dissolving, so did the primary condition that produced his best work. When the time came when it was safe to reveal all (or almost all), Williams had little to struggle against, and the plays he wrote in that period suffered. Williams indeed was afflicted with a measure of the homophobia his critics accuse him of harboring. More important than the homophobia, however, is the fact that he fought against it, and that this struggle produced plays that are part of America's most enduring dramatic legacy.

III

It is probable that during his second stay in New Orleans, Williams also wrote the strange and strung-out one act *Auto-Da-Fé*. For inspiration, he returned to 722 Toulouse. As *Lord Byron's Love Letter* is a metaphor for a playwright divided against himself, in which Williams dramatized the inner conflict between the need to conceal and the urge to reveal, in *Auto-Da-Fé* he borrowed the name of his friend Eloi Bordelon and divides it between two characters with warring needs. It is with this play—wild, passionate, slightly mad—that one begins to find the authentic Williams.

The play takes place on the porch of an old frame house in the Vieux Carré. Even in his stage directions, which often contain important clues to Williams's themes and shouldn't be overlooked, Williams shows us how intensely the battle between transgressive urges and the desire to suppress them is being waged within him. As Williams describes them, the street and the Vieux Carré itself are not quaint tourist attractions. *There is an effect of sinister antiquity in the setting,* he writes in the set description, *even the flowers suggesting the richness of decay. Not far off on Bourbon Street the lurid procession of bars and hot spots throws out distance-muted strains of the jukeorgans and occasional shouts of laughter.* The sensual stench of New Orleans is everywhere; it permeates the atmosphere. It seeps out of the bars and hotspots through the streets into the private bedrooms of once-respectable homes. Indeed, the description suggests that the dark places suffused with hot music and sweaty sex are marching inexorably toward all those of clean mind and heart, and will overwhelm them. The description also suggests—*sinister antiquity*—the atavistic; that these feelings are very old (like The Old Woman) and very strong, threatening to devour the new, the "civilized."[22]

There are two characters, mother and son. Eloi, in his late thirties, is frail, gaunt, ascetic. His eyes burn with an inner fever. Williams doesn't describe his mother, Mme. Duvenet, except to say that both characters are fanatics. His interest is in the arresting Eloi—high-strung, hyper-sensitive, hypochondriacal, paranoid, puritanical, and deeply repressed. He complains to his mother that their border—Miss Bordelon—goes into his room when he's away and roots through his things. He suspects that she is a hired investigator who listens through the wall at night to hear him when he talks in his sleep. The fact that someone might be entering his room alarms him immensely; he even forbids his mother entrance to clean it. "A person would think," she says, "that you were concealing something." "What would I have to conceal?" he demands. She replies, "Nothing that I can imagine" (132).

This is no doubt true. Whatever is on Eloi's mind *is* beyond the realm of his mother's experience and imagination. It seems that while working at the post office, Eloi came across a photograph that, he says, fell out of an unsealed envelope. It's a pornographic picture of two naked figures which, according to Eloi, passes all description. We don't learn more about the nature of the photo (repeating the pattern set in *Lord Byron's Love Letter*), but we can guess. It was addressed, Eloi says, to "One of those—opulent—antique dealers on—Royal . . ." and was sent by a university student (145–6). Rather than report this crime to the authorities, however, Eloi has conducted his own "investigation." He has visited the sender in the privacy of his dormitory room (the student is a male—or else the two could not have talked in private in a university dormitory in 1941). The interview was now three days ago, yet Eloi has taken no further action. Indeed, he still possesses the photograph; he even carries it with him. He is torn as to what to do. Mme. Duvenet suggests he destroy it, since, having hesitated so long without informing the authorities, he's likely to appear culpable himself. The paranoid Eloi urges his mother to lower her voice: the unseen Miss Bordelon obviously has been hired by the antiques dealer to spy on him and is eavesdropping this very minute. His mother insists he burn the picture immediately, but in his extreme nervousness, he burns himself with the match instead. Rushing into the house he locks the screen door behind him. Mme. Duvenet hears him confronting Miss Bordelon. Eloi raises his voice violently and Miss Bordelon cries out in fear. Something metallic smashes against the wall. Flames burst through the inner rooms, and the last we see is Mme. Duvenet staggering down the porch steps shouting, "Fire! Fire! The house is on fire, on fire! The house is on fire!" (151).

What can we notice about this melodramatic little play, which in its emotional intensity, luridness and violence, prefigures so many of the plays to come? One fact is that despite his age—Eloi is in his late thirties—he has not left home. He insists that he hates the obscene atmosphere of the French Quarter, that it makes him physically ill, yet he won't move away. What keeps him here? Next, there is the intense delight with which Eloi, like so many puritans, talks about filth. Everything comes back to dirt. From the dust and disorder of his room (which he refuses to let his mother clean) to the very air around them, the city, he tells his mother, is permeated with filth.

> This fetid old swamp we live in, the Vieux Carré! Every imaginable kind of degeneracy springs up here, not at arm's length, even, but right in our presence! [. . .] This is the primary lesion, the—focal infection, the—chancre! In medical language, it spreads by—metastasis! It creeps through the capillaries and into the main blood vessels. From there it is spread all through the surrounding tissue! Finally nothing is left outside the decay! (134–5)

Eloi's solution is equally fanatical. The city ought to be burned to the ground, "purified with fire" (136).

Eloi is torn between his hysterical sexual obsession and the guilt it induces, which is represented by Miss Bordelon, who, he's convinced, waits and listens inside. She is his guilty conscience, split off from him yet part of him, just as Williams took his friend Eloi Bordelon and figuratively split him in half to represent the playwright's own ambivalent self.

In the end, Eloi cannot reconcile himself to his homosexuality, and yet neither can he leave it behind him: If he could, he would have left the Quarter long ago. He is a slave to his obsession, and it also deeply shames him. The only way out for him is death, and death of the most painful sort.

In many senses, Eloi is different from the gay characters Williams would create later. He is, for example, unable to cut himself loose, to flee—which is perhaps the archetypical action of almost all of the mature Williams's major characters. He is stuck, unable to go back into the house or stay on the porch. He shares this trait with his neighbor in *Lord Byron's Love Letter*, Ariadne. Also past her youth, she remains with her grandmother, unable to strike out on her own. In later plays, Williams will give us birds—flocks of nightingales and canaries—that symbolize an escape of a very specific sort. The only escape from suffocation or middle-class stuffiness, in *Auto-Da-Fé* is a fiery death. If, however, there was to be no earthly deliverance from the terrifying depths of Williams's feelings, in *Auto-Da-Fé* there was at least a new willingness to admit their existence. This is a major step forward.

What is still missing is another kind of escape: an escape from the conventions of theatrical realism. As Williams became truer to his promise to be wild, passionate, and mad, he would recognize that realism was too small a vision to contain his characters' increasingly out-sized emotions. The full-blown expressionism of *Orpheus Descending*, which also ends in fire and death, is yet to come.

Hysterical, hypochondriacal, filled beyond relief with loathing, Eloi is a gay activist's nightmare. He is hardly fit for presentation to impressionable young gay men and women as a role model. But then, the world from which both Eloi and Tennessee Williams came was hardly amenable to the gay role models or "positive images" that some critics feel is the one true test of a good gay playwright. Indeed, beginning in the late 1980s, gay critics would excoriate Williams in terms as harsh as those of the homo- and sexphobic critics of the 1950s and 60s for being unable, in their opinions, to portray gay characters in an acceptably "positive" light. The charge is false. With the exception of The Queen, one of the prison inmates in *Not About Nightingales* and, like the others in that play, little more than a type, Eloi is probably Williams's earliest gay character, dating from a time in the playwright's life when he, too, was struggling with questions of sexual identity and self-worth. But Eloi is only a starting point. By the time his career was over, Williams had moved from gay characters who were marginal or offstage, to characters who were out and at the center of his plays. It was a process that began long before Stonewall, as the rest of this book will show. For the moment, it is enough to point out a similarity between Eloi and Mme. Duvenet and another son and mother in the Williams canon.

In Mme. Duvenet's hovering, hectoring, and instructions on how and what to eat, her faith in all things spiritual, and the endless bickering with her son, one clearly hears a rehearsal for *The Glass Menagerie*. Mme. Duvenet is an easily recognizable prototype of Amanda, right down to her mixture of canny knowingness and simple naïveté. And while Eloi is far more disturbed and neurasthenic than Tom Wingfield, they share a penchant for secrets. "Oh, I could tell you many things to make you sleepless!" Tom will tell Amanda in a moment of anger. And in a quieter moment, as Tom obliquely tries to explain himself to his mother, he says, "You say there's so much in your heart that you can't describe to me. That's true of me, too. There's so much in my heart that I can't describe to *you!*" What Tom holds in his heart that he can't describe to his mother and the thing that makes Eloi a stranger in his own house, are one and the same.[23]

Eloi is not nearly as reconciled to his homosexuality as Tom will be, but then Williams, at this point, wasn't nearly as accepting of his as he would be a few years hence when writing *The Glass Menagerie*. While in *Auto-Da-Fé*, Williams either cannot or will not tell us more than Eloi tells his mother; in the new play that was shortly to grow in his mind, he would find a way to reveal to us what Tom conceals from Amanda. There is, then, a direct path from the mightily self-loathing Eloi to the more adjusted Tom. Williams was embarking on a path that eventually would lead him to The Writer in *Vieux Carré* and August in *Something Cloudy, Something Clear.*

IV

In addition to grappling with his inner demons and with the generalized, invisible homophobia in the society around him, Williams would soon face, along with every other American gay man and lesbian, homophobia in its institutionalized, very visible form. There is scant evidence suggesting that Williams was directly affected by the American military's successful effort to place homophobia on an official, government-sanctioned footing (he would be classified 4-F due to his poor eyesight). No homosexual, however, could escape the indirect effects of the prejudice about to sweep the country in the form of raids on gay bars, and newspaper and magazine reports (initiated by the military) on how unfit homosexuals were to join the fight against Fascism.

At about the time Williams was classified 4-F, Army psychiatrists were developing rationales and screening processes to ferret out gay men before they could slip unnoticed into the armed forces. The doctors described homosexuals as exhibiting one of three levels of mental illness: psychopaths, who were sexual perverts; paranoid personalities who suffered from homosexual panic; and schizoid personalities who displayed homosexual symptoms. Such was military thinking about gay men (and women, as well). In time, the stereotypes behind these categories filtered into the general public's thinking about gay men: They were sick sissies who were mentally unfit to serve. These attitudes, meant to be sympathetic and understanding, did not exist in a vacuum. The armed forces' psychiatrists considered them to be enlightened advancements over the general attitude of Americans who considered homosexuals to be criminals. A victim of mental illness, after all, was not responsible for his reprehensible actions.[24]

At the beginning of 1942, as Williams was about to abandon New Orleans a second time, the armed forces' antihomosexual screening standards were tightened. New army mobilization regulations included a paragraph called "Sexual Perversions": "Persons habitually or occasionally engaged in homosexual or other perverse sexual practices," it said, were "unsuitable for military service." It listed three telltale signs of homosexuality and clarified procedures for rejecting gay draftees: "feminine bodily characteristics," "effeminacy in dress and manner," and a "patulous [expanded] rectum." All of these "telltale" signs "linked homosexuality with effeminancy or sexually 'passive' anal intercourse and ignored gay men who were masculine or 'active' in anal intercourse." The paragraph was written by a rising star in the young American psychoanalytic movement, Lawrence Kubie. In 1957, he would become Williams's analyst.[25]

Defining who was homosexual in order to exclude them from service presented the military with a new problem: Any man who wanted to escape the draft merely had to declare himself a homosexual. To combat this tactic, high-ranking officers created and magnified "a widespread revulsion towards homosexuality both inside and outside the military." While military psychiatrists were engaged in what they considered a liberal reform of attitudes toward homosexuals, other officers launched a hate campaign that relied on preserving them: If a heterosexual chose to weasel out of his patriotic duty by masquerading as queer, he would pay a heavy price. Newspapers and magazines were quick to aid the military's efforts by publicizing their methods for weeding out homosexuals as undesirables. Repeatedly, they parroted the psychiatrists' links between homosexuality, perversion, and mental illness. In a full-page story in the *Washington Sunday Star,* a reporter wrote that Navy psychiatrists "will be on the lookout for any number of mental illnesses or deficiencies that would make a recruit a misfit." Among those "misfits," he wrote, were homosexuals. As historian Allan Bérubé writes, "Military unfitness was one of the few contexts in which the popular media discussed homosexuality at all during the war."[26]

Screening often took the form of an interview at the recruitment site. Potential inductees were asked a battery of questions, written or oral, often in rapid-fire sequence, to determine just how "normal" they were. These questions might be about the candidate's sexual habits or whether or not he liked girls. A wrong ("I don't like them") or even hesitant answer could lead to rejection and the classification "homosexualism-overt," a classification that became a permanent part of the man's record. Gay men who made it through the inspection but were at some point in their service discovered to be gay were given medical discharges, and the label "homosexual" was entered on their records.

Not long after Williams arrived in New York in 1942, another young playwright, Donald Vining, recorded in his diary his experience at a Pennsylvania induction center. The examining officer asked him if he got along with women. Although he expected such a question, and indeed had been searching for an opening to admit his homosexuality, Vining stuttered that he didn't associate with them much. That seemed to be sufficient. He received the classification "Sui generistic *H* overt," which was re-written "homosexualism-overt" when the classification officers couldn't understand the examining doctor's terminology.[27]

Resident aliens also were subject to the draft. One young Englishman who had come to America in 1939 on a scholarship to Yale decided to volunteer his services. In England, he'd been a conscientious objector; but by December 1942, he'd begun to feel differently about his personal responsibility and the war. He had recently received his Ph.D. from Yale, had taught at UCLA, and had just obtained a teaching job at a new experimental college in the South. Nonetheless, in that month, he presented himself for induction at the Park Avenue Armory in New York City. Clad only in his underwear, standing before a psychiatrist in a three-piece suit and requisite German accent, he was unprepared for the questions he was asked: "Do you like women?" and "Has the sight of a member of your own sex ever given you an erection?" Giving no thought to the consequences of a truthful answer, he responded honestly. The psychiatrist frowned. He passed the young man along to some uniformed officers. One was particularly aggressive, seemingly bent on trying to discover whether this now-befuddled, increasingly alarmed young scholar was lying to evade the draft. He demanded to know where the young man found sexual partners and what he did with them. Another officer was subtler. He asked the would-be soldier if the sight of a handsome young man ever gave him an erection, placing his arm around his shoulders as he did so. The young man managed to say, "Not usually." The comforting officer called him by his first name. "I'm sorry, Eric," he said. "I don't think the Army can use you at present." Eric put on his clothes and was escorted out. He had to pass through a military checkpoint to be released. The clerk there found his name on a list, putting next to it a capital H. The clerk's friends hovered around to see, and, spying the letter, shouted, "Homo! Another homo!" loud enough for others nearby to hear plainly. The young man staggered out onto the street, and had to lean against a wall for support. Because he was a homosexual (albeit one who was scheduled to be married the following week), Eric Bentley's military career was over before it began. He returned to his teaching job at Black Mountain College, where he introduced students, faculty, and townspeople to the plays of Bertolt Brecht.[28]

Being rejected for service due to "sexual abnormality" could expose a man to danger. His record, which included the reason for rejection, was sent to his local draft board, where it was not necessarily treated confidentially. Under the 1940 Selective Service Act, employers had the right to examine an applicant's draft record as a condition of employment. In the years prior to the war, if homosexuality was widely condemned, it was done tacitly, with the sort of silence such shameful behavior was thought to deserve. After Pearl Harbor, as the military's manpower needs increased, the condemnation of homosexuality became part of the public discourse. Even gay men would condemn their brethren who declared their homosexuality in order to be exempt from service.[29]

Although the Shore Patrol and Military Police's increased crack-downs on the bars put something of a damper on gay life in New York, Williams found many opportunities for sex, and he was rarely slow to take advantage. Indeed, it was clear to those who knew him then that Williams, although he was looking for love, didn't mind settling for sex. Donald Windham recalled that Williams's attitude toward sex "was so enthusiastic that he considered 'tender' feelings to be involved when his partner was someone known rather than a total stranger." One Saturday night in March 1942, he brought a group of merchant marines back to the apartment where he was staying. Such an event was ordinary enough that it rated only a laconic mention in his journal: "Today—I am dull but relatively at ease. Could not do any good writing. Last night the apt. was filled with merchant marines—this A.M. F's electric razor was missing. Which is *not* a *non sequitur.*" "F," Williams's host, the painter Fritz Bultman, was not amused, and the next day Williams found himself evicted.[30]

Movie houses were especially popular places in which to pick up men, and even have sex, and both civilians and servicemen cruised them regularly. Donald Vining, now working as a salesman at B. Altman's while pursuing his playwriting career, was a frequent visitor. At first, he found the hunt difficult to get used to. "At 7:30 I walked up to 42nd Street," he wrote in his diary. "Again I was amazed at the way men came in and then left in about five minutes. The sailors were cleaned out of there by 8:30. I still haven't gotten the technique and I still get engrossed in the pictures and forget what my true mission is." In short order, he learned to focus his attention in the proper place and recorded in his diaries more than one successful visit to the movie theatres in Times Square. Eventually, however, civilian and military police cracked down on these places. By 1945, after returning from a lengthy stay in California, Vining would note in his diary, "All warned me against theatre cruising because of the plainclothesmen."[31]

Sex was not without danger from other sources. A man out cruising might pick up a trick he thought was "trade" (straight men who let gay men have sex with them, usually fellatio) only to be beaten up and robbed. Williams, in his hunt for sex and affection, was victimized at least once. One might think, given the level of self-hatred some critics accuse Williams of harboring, that such occasions would result in some expression of loathing. But his journals at the time show emotions of another sort. His response to an episode that occurred early in 1943 was complex. He expresses regret—not at being gay, but at being the target of another man's acute discomfort. He was staying at the St. George Hotel in Brooklyn Heights, and had been cruising the Promenade.

> This is the first time that anybody ever knocked me down and so I suppose it ought to be recorded. Unhappily I can't go into details. It was a case of guilt and shame in which I was relatively the innocent party, since I merely offered entertainment which was accepted with apparent gratitude until the untimely entrance of other parties. Feel a little sorrowful about [it]. So unnecessary. The sort of behavior pattern imposed by the conventional falsehoods.
>
> Donnie [Windham] comforted me when he arrived on the scene. Now he is upstairs with another party procured in the bar. Why do they strike us? What is our offense? We offer them a truth which they cannot bear to confess except in privacy and in the dark—a truth which is inherently as bright as the morning sun. He struck me because he did what I did and his friends discovered it. Yes, it hurt—inside. I do not know if I will be able to

sleep. But tomorrow I suppose the swollen face will be normal again and I will pick up the usual thread of life.[32]

The next day, he wrote:

> There was something incredibly tender and sad in the experience. So much of life at its most haunting and inexpressible. Not that I like being struck, I hated it, but the keenness of the emotional situation, the material for art—these gave a tone of richness to it which makes the affair unforgettable among many that melt out of sight.[33]

What Williams sensed was not only the violence—which he would reflect again and again in his work—but what lay beneath it: the lack of love. In the end, this lack of love would interest him more as an artist than would any protest against the violence. Not that he would be silent when it came to protesting violence and cruelty; indeed, his work as a whole can be taken as a protest against it, and he would be vocal in his position for gay rights as he grew older. But his instinct was for something larger than gay rights. He was concerned for all of those who were victimized, brutalized by the rapaciousness of the world; he responded to an existential loneliness and isolation, one that he certainly sensed in himself.

Whatever else he concealed from himself, Williams was constantly aware of loneliness's tug and pull. It was, he wrote in his *Memoirs,* the major theme of his writing and his greatest affliction, "the affliction of loneliness that follows me like my shadow, a very ponderous shadow too heavy to drag after me all of my days and nights."[34]

Unfortunately, ending loneliness meant confronting those very powerful emotions and desires from which he sought to protect himself, emotions that he would siphon off into his work. And even there, for a very long time, he managed to conceal them:

> I remember when I first came to N.Y. . . . 3 seasons ago—I kept looking for "the big, important thing" to happen. Well, that summer I had K[ip]. It nearly killed me . . . But a life in a small place, with simple, honest relationships (friends) *and* some sexual partner or outlet accessible is what I should seek out for myself.[35]

In his work, the struggle between loneliness and the fear of intense feelings of love would be expressed in dramatic situations where the stakes would literally become life and death. In *Orpheus Descending,* it is Val's irresistible love for Lady that fatally leads him to stay in Two River County, and be burned alive by the remnants of Jabe Torrance's gang. Val is a loner, suspicious of those seeking attachments—especially emotional attachments. The other side of that coin is the tremendous need Williams tried to shield himself from, the feeling voiced at the end of Act Two when Lady exclaims to Val, in capital letters and exclamation marks, "NO, NO, DON'T GO . . . I NEED YOU!!! TO LIVE . . . TO GO ON LIVING!!!"[36]

Concealing versus revealing; staying versus fleeing; the presence of deeply felt emotions that must be kept at arm's length. Powerful conflicts, all, and all would be powerfully presented in the full-length play he'd write upon arriving, in May 1943, in California.

V

When will the sleeping tiger stir
among the jungles of the heart?
I seem to hear the sound of her
gentle breathing in the dark.

O you that are deceived by this
apparent innocence, take care!
You know that storms are presaged by
such trembling stillness in the air.

And all that breathe have in their breast
capacity for certain flame,
Domesticated cats are merely
beasts pretending to be tame.

Not for the pelt but for the passion
would I track that tiger down,
to dwell with her more dangerously
beyond the lighted streets of town!

(1940)[37]

Williams arrived in Hollywood courtesy of his agent Audrey Wood. She had gotten him a position as a staff writer at MGM. "Sold to Hollywood," he wrote in his notebook. It was a good deal. His first assignment was to re-write a screenplay for Lana Turner called *Marriage Is a Private Affair.* The material hardly engaged his interest, however, and he spent large portions of his days working on the scenario of a piece called *The Gentleman Caller.* Whether it was a screenplay or a play he couldn't decide. By the end of June, he was taken off the Turner picture, but, much to his pleasure, he remained on the payroll. He was laid off for six weeks starting in August, but then, according to the terms of his contract, was reinstated for the balance of his six-month stint. Still he had no official assignment, so work on *The Gentleman Caller* continued.[38]

At night he was free to roam the area around Santa Monica, where he lived in an apartment building on Ocean Avenue. Once again, there were soldiers and sailors in abundance, and Williams wasted no time. He rarely lacked for casual sex:

> The Palisades were full of young servicemen, positively infested with them, I'd say, and when I'd driven by one [on a motor scooter he drove perilously around Santa Monica] who appealed to my lascivious glance, I would turn the bike about and draw up alongside him to join him in spurious enjoyment of the view.
>
> Presently I would strike a match for a cigarette. If the match-light confirmed my first impression of his charms, I would mention that I had a pad only a few blocks away, and he would often accept the invitation. If the first one or two were not to my satisfaction, I would go out for a third.[39]

By day, first at his MGM office and later in the apartment on Ocean Avenue, Williams was writing what would soon become *The Glass Menagerie.* At night, on the Palisades, at the movies, and even on the bus ride home when he would contrive to have his knee jostle one of a likely partner, he was creating the pivotal material that comprises the play's hidden heart.

No doubt Williams's sense of loneliness was exacerbated by the alienation felt by a gay man born in Mississippi in 1911 who came to maturity during the years of America's increasingly institutionalized homophobia. But the roots of loneliness, like those of sexuality, go deeper than can be explained by social and political paradigms. They lie in Williams's experience of his family, in the feelings aroused by a fa-

Not a word of dialogue has been spoken before Williams gives us a very specific picture: Although some of the world is "rather dim and poetic," the play's environment is one of desperation. It is also one of suffocating closeness and stultification. The situation in which much of the middle class finds itself is one of gray dronesmanship; their homes are hives and "warty growths." We don't need knowledge of the plays to come to understand Williams's attitude toward this world and the conformity it imposes on its citizens. It is a world to escape. The description of the apartment's interior adds little hope for the prospects of those who inhabit this world:

> *Nearest the audience is the living room, which also serves as a sleeping room for Laura, the sofa unfolding to make her bed. Just beyond, separated from the living room by a wide arch or second proscenium with transparent faded portieres (or second curtain), is the dining room. In an old-fashioned whatnot in the living room are seen scores of transparent glass animals. A blown-up photograph of the father hangs on the wall of the living room, to the left of the archway. It is the face of a very handsome young man in a doughboy's First World War cap. He is gallantly smiling, ineluctably smiling, as if to say, "I will be smiling forever."*

> *Also hanging on the wall, near the photograph, are a typewriter keyboard chart and a Gregg shorthand diagram. An upright typewriter on a small table stands beneath the charts.* (143–4)

There are two essential circumstances ("given circumstances" in Stanislavskian terms) we can deduce from Williams's set description: *there is very little money and there is very little space—meaning very little privacy.* When desperate people live in such circumstances, their desperation becomes all the more intense. The emotional atmosphere is further strained when two of those living in such a place are young people who, given such close quarters, must always be under the scrutiny of a demanding parent.

Williams, however, has provided a few ways out of this desperate world. One is exemplified by the young man who is smiling forever. He smiles because he up and left with no more than a two-word postcard: "Hello—Goodbye!" There is, of course, the "whatnot" with its glass creatures, from which Laura has fashioned quite a successful escape into herself. The typewriter keyboard and shorthand charts were meant, courtesy of Amanda, to provide Laura with another escape, a route into financial independence. Laura, however, has resisted these efforts as successfully as she has all others. As for the typewriter, Amanda meant it to be another way out for Laura. Instead, it has become one for her son: His writing takes him far away from the desperate conditions of his home life. Just as Laura's ability to withstand all of her mother's plans to "socialize" her indicates the strength of her will, so Tom's ability to write—to escape within himself—at a kitchen table with absolutely no privacy suggests his. The difference, here, between Tom and Laura, is that while Laura's will power is limited to escaping into an interior world, Tom has a second outlet for escape, outside himself.

Unlike Eloi, Tom will succeed in escaping, if only in part. To use another term from Stanislavski, Tom's overall objective—that goal that he wants to achieve above all, and toward which most of his actions are devoted—is to escape. If one wants to escape, then one is fleeing one thing for another. What Tom is escaping from is the prison of the apartment, dominated by his mother and sister and devoid of privacy, and the sterility of work in the celotex interior of Continental Shoemakers, where he slaves daily. What is he escaping toward?

That homosexuality is present in *The Glass Menagerie* is not a new critical idea. But for some critics, that homosexuality is purely metaphorical and abstract, existing as a literary inference, rather than functioning as a vital part of the world of the play. The critic Mark Lilly sees Laura's lameness as a metaphor for Williams's view of homosexuality (and not, clearly, a very flattering one). "First," Lilly writes, "it is seen as a disability that actually restricts sexual fulfillment." Indeed, it is particularly unhealthy because it causes Laura to withdraw from the world. Jim, her one gentleman caller, views this disability, according to Lilly, as "quaint and unhealthy," and as a representative of the "world of 'normality,'" he tries to entice her away from it into his own "normal" orbit. All in all, Lilly finds Laura to be "unhappy" and "unfulfilled," although there is nothing in the play to suggest that this is a condition that will last beyond the time it takes her to recover from the gentleman caller's headlong flight in the face of her abnormality. Still, her menagerie, which, according to Lilly, is another symbol for homosexuality, possesses a positive potential:

" . . . the possibility we all possess of living in an individual reality other than and hostile to the reality foisted on us by society . . ." But, he continues, "this individual reality will probably be found to be vulnerable and fragile."[47]

Where does this leave the actor or director? Nowhere, because while such metaphors and inferences can illuminate novels, stories, or poems, they don't go very far in explaining the behavior of dramatic characters who, on a stage, are represented by human beings who must act as human beings always do: in response to specific desires and circumstances. Let us state clearly, then, that the homosexuality in *The Glass Menagerie* resides in Tom, and that it is not metaphorical or abstract. Tom is gay.

He is also cagey. He is an embodiment of Williams's intense conflict between the urge to conceal and the concomitant desire to reveal. The opening paragraph of his first speech tells us as much.

> Yes, I have tricks in my pocket, I have things up my sleeve. But I am the opposite of a stage magician. He gives you illusion that has the appearance of truth. I give you truth in the pleasant disguise of illusion. (144)

There is truth here, in other words, but it is disguised. He continues:

> To begin with, I turn back time. I reverse it to that quaint period, the thirties, when the huge middle class of America was matriculating in a school for the blind. Their eyes had failed them or they had failed their eyes, and so they were having their fingers pressed forcibly down on the fiery Braille alphabet of a dissolving economy.
>
> In Spain there was revolution. Here there was only shouting and confusion. In Spain there was Guernica. Here there were disturbances of labor, sometimes pretty violent, in otherwise peaceful cities such as Chicago, Cleveland, Saint Louis . . . (144–5)

What the middle class could not see, on one level, is a way out of an economic disaster that none had foreseen. On another level, the message is broader: they could not see, or refused to see, what is there to be seen if one knows how and where to look. "In Spain there was revolution. Here there was only shouting and confusion." In a distant revolution, issues are clear-cut, there is Right and Wrong. Close up, at home, the issues are buried, and nothing is completely clear: there is "only shouting and confusion."

The play, Tom tells us next, "is dimly lighted" (145). That is, it takes place neither in total darkness nor in the blinding light of day. There are layers to be peeled back, but not discarded. The layers and the shadows are as important as the secret they conceal; the play's darkness can't be removed without destroying the play. What is hidden is only to be glimpsed, not brought fully into the open. Those who can read the signs, whose eyes do not fail them, will see Tom's gayness; those who cannot will not. Light and dark, truth and illusion, are all of equal value in *The Glass Menagerie*. This will displease those critics who consider good only gayness that is completely visible and proud. But then, their yearning for conformity to a localized standard places them in that hive of society that avoids differentiation and would prefer the gay world to function according to their own sort of automatism.

The play moves on to conflict immediately, as Amanda instructs Tom how to eat. It is one we've already witnessed between one gay son and a mother who cannot see. The first mother spoke this way: "Chew your food, don't gulp it. Eat like a human being and not like a dog!" (139–40) Amanda expands operatically on Mme. Duvenet's admonition to Eloi. Indeed, in many respects, *The Glass Menagerie* is an expansion on and development of the relationship between mother and son in *Auto-Da-Fé*.

Scene Three is largely devoted to another intensely emotional confrontation (for a play that is supposedly so sweet and nostalgic, *The Glass Menagerie* has a lot of these). Tom and Amanda quarrel violently for five pages because Amanda has interrupted Tom's writing. She has invaded his privacy by peering over his shoulder as he types. He is furious at this trespass, at her attempt to see what he is writing. Perhaps the trespass alone is enough to set him off, or perhaps he specifically wants to keep his work secret from her. In turn, he may want to keep the work secret because its contents are gay—or just intensely private. All of these are choices to be made by actor and director. Their specificity begins to build up a densely detailed secret life, the specifics known only to these two (or perhaps only to the actor). But if the choices they make regarding that life, unknown to the audience though they may be, are sufficiently specific, they will cast a shadow in Tom's wake that an audience *will* sense.

Individual members of the audience will reach their own conclusions as to the nature of that secret life. Those who have eyes to see will see. Those who do not will see . . . something else. This is not evidence of the playwright's self-loathing; it is Williams's artistic solution for surviving the intensity of his inner life. That very desperate need for survival has been turned to artistic use: it is funneled into characters who adopt that desperation as their own, and its subtextuality gives *The Glass Menagerie* a power and an inner turmoil so central that to ignore them is to misunderstand the play.

The actor and director have more choices to make about Tom's private life as the scene continues. In a fury, Tom leaves the house bound for the movies, but not before Amanda demonstrates very clearly that although she may be foolish, she is not stupid. Unlike her predecessor, Mme. Duvenet, she has some idea of what is going on around her:

> I think you've been doing things that you're ashamed of. That's why you act like this. I don't believe that you go every night to the movies. Nobody goes to the movies night after night. Nobody in their right minds goes to the movies as often as you pretend to. People don't go to the movies at nearly midnight, and movies don't let out at two a.m. Come in stumbling. Muttering to yourself like a maniac! (163)

Tom taunts Amanda with a made-up story about his "secret life" when she demands, again, to know where he is going. Echoing Eloi ("You just don't know. [. . .] There's lots of things that you don't know about, Mother" [139]), he tells his mother that she has no idea about who he really is, what he really does.

> I'm going to opium dens! Yes, opium dens, dens of vice and criminals' hangouts, Mother. I've joined the Hogan gang, I'm a hired assassin, I carry a tommy gun in a violin case! I run a string of cat houses in the Valley! They call me Killer, Killer Wingfield, I'm leading a double-life, a simple, honest warehouse worker by day, by night a dynamic *czar* of the *underworld, Mother.* I go to gambling casinos, I spin away fortunes on a roulette table! I wear a patch over one eye and a false moustache, sometimes I put on green whiskers. On those occasions they call me—*El Diablo!* Oh I could tell you many things to make you sleepless! My enemies plan to dynamite this place. They're going to blow us all sky-high some night! I'll be glad, very happy, and so will you! You'll go up, up on a broomstick, over

Blue Mountain with your seventeen gentlemen callers! You ugly—bab-
bling old—*witch*. . . . (164)

There is fury here, but there is something else, as well. One senses that
Tom *enjoys* not only taunting his mother, but dropping hints about the
secret life that he actually does lead. Of course, an actor or director could
choose to believe that nothing in this speech is in any sense true, that it is
all metaphorical, made up. After all, Tom is a writer who lives a large part
of his life in his imagination. But also as a writer, he uses his own life for
material. The real fury and anger come from his real secret life, and the
enjoyment he takes is that of a writer dropping hints that his "reader,"
Amanda, won't see. What is the nature of the double life he describes in
images that come straight from the movies? What are the things that he
could tell his mother—what can any young gay man tell his mother—to
make her sleepless? That he has a girl friend stashed away somewhere?

The key to understanding that Tom is gay comes in the next scene. It
lies not only in what Tom says but also in what he can't say. It lies in the
circumstances, and in the gift he gives to Laura.

The first part of Scene Four occurs several hours later. It is five o'clock
in the morning. Laura, sleepless with worry that this time her brother
will never come home, is awake. As Tom fishes for his door key, we know
that in fact he has been to the movies: a shower of ticket stubs cascades
from his pocket. But, as Amanda correctly observes, no movie theatre is
open so late. Tom tells Laura he has gotten drunk through the good
graces of Malvolio the Magician, the star of the movie theatre stage show
who called on Tom as a volunteer to drink the water that he turned to
whiskey. But consider the clock: If Malvolio performed between show-
times of the feature, Tom probably got drunk somewhere between nine
and ten. Eight hours later, he is still so drunk that he has difficulty climbing the fire escape steps, and then drops his key through a crack. An actor
and director faced with this scene have to contend not with metaphors of
gayness but with the reality of behavior. If Tom has been to the movies,
he has been somewhere else besides.

Williams gives us a clue. Tom carries a noisemaker. Where did he get
it? Here, director and actor must make another choice, one that is not
only logical but that provides specific answers to questions about Tom's

off-stage life. Therefore, one might conclude that the noisemaker comes from the same place where Tom has spent much of the last eight hours drinking: a bar, perhaps, or a party.

This raises another question. Why doesn't Tom tell this to Laura? They have a close and deep relationship: so deep that, as far as Tom will run from his family, he will always see his sister's face in every darkened shop window, in every bit of rainbow-colored glass. Nevertheless, he can say no more to her of his night out than he might to his mother:

> There was a big stage show! The headliner on this stage show was Malvolio the Magician. He performed wonderful tricks, many of them, such as pouring water back and forth between pitchers. First it turned to wine and then it turned to beer and then it turned to whiskey. I know it was whiskey it finally turned into because he needed somebody to come up out of the audience to help him, and I came up—both shows! It was Kentucky Straight Bourbon. A very generous fellow, he gave souvenirs. [*He pulls from his back pocket a shimmering rainbow-colored scarf.*] He gave me this. You can have it, Laura. You wave it over a canary cage and you get a bowl of goldfish. You wave it over the goldfish bowl and they fly away canaries. . . . But the wonderfullest trick of all was the coffin trick. We nailed him into a coffin and he got out of the coffin without removing one nail. [*He has come inside.*] There is a trick that would come in handy for me— get me out of this two-by-four situation! (167)

He says nothing to Laura about where he's been for the last several hours, where he got that noisemaker, and where he got so drunk (and she doesn't ask). If he had merely been drinking at a bar, or met a girlfriend, why couldn't he confide in Laura? What can't he say to her? What could he tell her that might make *her* sleepless, that he fears she couldn't understand?

What he does describe in detail is Malvolio's trick with the rainbow-colored scarf. "You wave it over a canary cage and you get a bowl of goldfish. You wave it over the goldfish bowl and they fly away canaries." Tom's overall objective, I've said, is to escape—from the goldfish bowl of the life he shares with his mother and sister, and from the prison of work, into a world where he can be his own authentic self, where he can express himself fully as an artist and as a man, a sexual being. Malvolio's scarf is the magic agent that effects this transformation. It is perfectly possible, of course, that Tom received the scarf from Malvolio at the movies just as he described. It is also possible that the scarf, which Tom makes a gift of to Laura, is more important to him than that. It is a gift representing the freedom that Tom wishes for himself—and for his sister.

It is quite possible, in other words, that the scarf was a gift to Tom from someone whom he cannot talk about, who represents to him, in some vital way, freedom. It may well be a gift from a lover (that lover could be Malvolio the Magician!). This, in turn, has implications for the way in which Tom handles the handkerchief, the way in which he hands it to his sister, and what the giving of it means. If, indeed, it was a gift to him from a lover who turns goldfish into canaries (an image that will also be associated with homosexuality in *Camino Real* and escape in general in play after play), then it is a very precious thing. It is no less precious a gift when he gives it to Laura: The freedom that it represents for him is what he wishes for her. He cannot be more explicit with her; he can only present her with this most valuable of wishes. It is a quiet but tremendously potent moment between brother and sister, not one to be tossed away.

Tom drops another hint about his hidden homosexuality in the second part of Scene Four, the next morning. Gruffly, he apologizes to his mother. Worried (with good reason) that her son will follow in the disappearing footsteps of her husband, Amanda takes the opportunity to confide in him—and also to probe for some hint to the hidden areas of his life:

> Amanda: Laura says that you hate the apartment and that you go out nights to get away from it! Is that true, Tom?
> Tom: No. You say there's so much in your heart that you can't describe to me. That's true of me, too. There's so much in my heart that I can't describe to *you!* So let's respect each other's—
> Amanda: But why—*why,* Tom—are you always so *restless?* Where do you *go* to, nights?
> Tom: I—go to the movies.
> Amanda: Why do you go to the movies so much, Tom?
> Tom: I go to the movies because—I like adventure. Adventure is something I don't have much of at work, so I go to the movies.
> Amanda: But, Tom, you go to the movies *entirely* too *much!*
> Tom: I like a lot of adventure.
> [*Amanda looks baffled, then hurt. As the familiar inquisition resumes, Tom becomes hard and impatient again. Amanda slips back into her querulous attitude toward him.*]
> [*Image on screen:* A sailing vessel with Jolly Roger.] (173)

The pauses that Williams indicates in Tom's responses indicate that he is searching for an evasive answer. He goes to—the movies for . . . adventure. Williams knew quite well why so many men went to the movie houses that lined 42nd Street in New York, and in Hollywood. He knew how men looked for adventure in the French Quarter, in Provincetown, in Mexico, on the Pacific Pallisades, even on darkened buses en route to Santa Monica. Tom, like Tennessee, likes a lot of adventure. Later, Williams would set stories of explicit gay sex, such as "The Mysteries of the Joy Rio" and "Hard Candy," in the darkened balconies of movie houses.

Tom keeps secret from Amanda much that is in his heart, but he doesn't keep it secret from us. We see it clearly in the image of the Jolly Roger, the image of a community of men set free from the conventions of a land-bound society.

Tom continues to shed dim light on his own inner life in his description, in Scene Five, of the Paradise Dance Hall across the alley from the apartment. Straight couples come to dance here, but also to make out in the alley. Dominating the ballroom is "[. . .] a large glass sphere that hung from the ceiling. It would turn slowly about and filter the dusk with delicate rainbow colors." Then, he makes the connection between the movies, rainbow-colored scarves, sex and escape explicit: "In Spain there was Guernica! But here there was only hot swing music and liquor, dance halls, bars, and movies, and sex that hung in the gloom like a chandelier and flooded the world with brief, deceptive rainbows . . ." (179). We should note that sex—sex between men—hangs not in the light, but in the gloom—it also floods the world with rainbows that are deceptive.

Moments later, Tom and his mother wish on the moon. Amanda asks what his wish was. His reply: "That's a secret." (180) The wish might be any number of things. It could be that thing he desires so deeply: an escape into the company of men.

Even when Tom's attention is on someone else, he reveals, through typical indirection, his true self. In Scene Six, he introduces us to the famous Gentleman Caller. His description is not merely reportorial:

> And so the following evening I brought Jim home to dinner. I had known Jim slightly in high school. In high school Jim was a hero. He had

tremendous Irish good nature and vitality with the scrubbed and polished look of white chinaware. He seemed to move in a continual spot-
light. He was a star in basketball, captain of the debating club, president
of the senior class and the glee club and he sang the male lead in the annual light operas. He was always running or bounding, never just walking. He seemed always at the point of defeating the law of gravity. (190)

There seems to be a part of Tom that wishes the Gentleman Caller was
calling on him.

There are so many hints about Tom's homosexuality scattered
throughout the play, if one has eyes to see them. But can Tom be played
explicitly as gay? No. That would violate the premise of *The Glass
Menagerie*. The play *is* gauze-covered, but what the gauze covers is the
desperation of intense sexual desires. Together, the desires and what covers them give the play its power, and would do the same for many of the
plays that followed.

Williams writes in his *Memoirs* that his goal in writing is to catch the
truth of a thing: "That goal is just somehow to capture the constantly
evanescent quality of existence. When I do that, then I have accomplished something, but I have done it, I think, relatively few times compared to the times I have attempted it. I don't have any sense of being a
fulfilled artist. And when I was writing *Menagerie*, I didn't know that I
was capturing it [. . .]" One might say that what Williams didn't know
he was capturing was the concealed and revealed "fact" of Tom's
gayness—the evanescent quality of *that* kind of existence.[48]

Williams needed his urge to conceal as deeply as he needed the one to
reveal. This was never a matter of self-loathing, but of surviving the intensity of his desire, not only for sex but for love. After the searing experience of Kip, Williams would not be so vulnerable again. This was not
good for his personal life, but it created the conditions that made the best
work of his artistic life possible.

The Glass Menagerie is an artistic triumph; for Williams, it was a
tremendous personal victory, as well. From the mechanical beginnings of
Lord Byron's Love Letter, in which he first employed the strategy of revealing and concealing, through the wilder, more personal, but not yet sufficiently controlled *Auto-Da-Fé*, Williams struggled to introduce into his
work subject matter that was not talked about in polite society, let alone

the commercial theatre of the 1930s and '40s. With *Glass Menagerie,* he came close to perfecting a technique that allowed him to integrate personal material with a theatrical structure in a way that enabled him not only to have the work produced, but to protect that part of his psyche that lacked the Williams family's "pioneering spirit."

When emotions such as those Williams felt while writing *The Glass Menagerie* must be expressed subtextually, they are often rendered all the stronger; when boundaries are established against expression, expression must find another way to break through. Igor Stravinsky had this in mind when he observed, "The more art is controlled, limited, worked over, the more it is free."[49] In the years to come, Williams would have to rely more on this strategy, not less.

TWO

The Time and World That I Live In

Police harassment, surveillance, and arrest; gay men incarcerated in violent wards; a government-sanctioned, systematic drive to hound homosexuals from their jobs: This was the atmosphere in which Williams wrote *Camino Real.* While problematic as a play, *Camino Real* is also an important document in the history of gay characters in American drama. At the height of a period unprecedented in its fear, paranoia, and homophobia, Tennessee Williams created an openly gay character for a play to be produced commercially on Broadway. Moreover, all of the damning comments of latter-day critics notwithstanding, only Tennessee Williams, among all his contemporaries, presented on a Broadway stage the image of an unashamed, democratic, and unvarnished gay man, expert in the sorts of sexual habits that seem to make recent critics (even the gay ones) as uncomfortable, even as homophobic, as any 1950s cop, politician, or commentator.

I

"Sometimes," Blanche says to Mitch in *A Streetcar Named Desire,* "there's God so quickly." Many gay men may have felt the same way on January 3, 1948, one month to the day after the play's Broadway opening. That day, *Sexual Behavior in the Human Male* first appeared in bookstores. Based on over 10,000 face-to-face interviews with American men and women and challenging nearly every widespread assumption about sexuality, the Kinsey Report, as it quickly became known, all 804 pages of it, became an instant best-seller. Within two weeks, over 185,000 copies were in print. By July, over 200,000 copies had been sold, and the publisher was running two presses around the clock to keep up with demand. The Kinsey Report quickly entered American pop folklore; it became the subject of adoring pop songs: "Ooh, Dr. Kinsey," and "The Kinsey Boogie" and "Thank You, Mr. Kinsey." All three became hits.[1]

Among Kinsey's principal findings was that male homosexual experiences were far more common than had been thought. Fifty percent of men interviewed reported they had had, at one time or another, an erotic response to another male. Thirty-seven percent said they had had at least one post-adolescent homosexual encounter that resulted in an orgasm; four percent said they had been exclusively homosexual throughout adulthood, while one in eight said that attraction to their own sex predominated in their sex lives for a period of a least three years. "Persons with homosexual histories are to be found in every age group, in every social level, in every conceivable occupation, in cities and on farms, and in the most remote areas of the country," Kinsey wrote. He also found that more than 90 percent of the interviewees had masturbated, 85 percent had engaged in premarital sex, between 30 and 45 percent had extramarital sex, and that 70 percent had patronized prostitutes. Kinsey summed up, "There is no American pattern of sexual behavior, but scores of patterns, each of which is confined to a particular segment of our society."[2]

Kinsey's findings led him to the conclusion that there was nothing the least abnormal about homosexuality. It was, he wrote, "an inherent physiologic capacity." Given the tremendous social and legal penalties for homosexuality in America, Kinsey concluded that in a sexually open, egalitarian society, his figures would have been even higher.[3]

Kinsey's statistics struck most homosexuals as good news: There were many more of them than anyone had ever suspected. Young people who were just discovering their sexual nature could take heart in the fact that they were not alone. "In effect," the historian John D'Emilio would write, "Kinsey's work gave an added push at a crucial time to the emergence of an urban gay subculture. Kinsey also provided ideological ammunition that lesbians and homosexuals might use once they began to fight for equality."[4]

In California, one of those homosexuals was in a moment of crisis. The years following the war were a period of terror for this married man and father. He feared for both his public life as a heterosexual and his private life, which was punctuated with furtive affairs with other men. The pretext of his life, his biographer would later write, was becoming unbearable, and the man himself said, "I was confronted by the horror of my own existence. I didn't know what to do."[5]

This was not entirely true. The man—who had been one of Kinsey's 10,000 interviewees—had been thinking about the need to transform gay men from an inchoate, isolated collection of alienated loners into an organized political minority group. He wrote later:

> The post-war reaction, the shutting down of open communication, was already of concern to many of us progressives. I knew the government was going to look for a new enemy. . . . [b]ut Blacks were beginning to organize and the horror of the Holocaust was too recent to put the Jews in this position. The natural scapegoat would be us, the Queers. . . . They—we—had to get started.[6]

The man was Harry Hay, and he was working out the beginnings of what would become the Mattachine Society, the first nationwide society of gay men.

It is worth looking at Hay for many reasons, not the least of which is to witness how difficult it could be, even for a man as progressive as Hay, to overcome the myths and stereotypes about gay men and women that were so prevalent in the post-war years, and that shaped the psyches of so many gays and lesbians, including Tennessee Williams. Hay's struggles with psychology and language, as well as his choice to make the Mattachine Society a secret one, help describe the social and political context in which Williams would create an openly gay character for *Camino Real* and then present the play as a commercial production on Broadway in 1953.

In August 1948, eight months following the premiere of *A Streetcar Named Desire*, Hay had begun formulating the notions that would lead to Mattachine. At first, he called the organization Bachelors Anonymous, and patterned it in significant ways after Alcoholics Anonymous. Even Hay, as radical a thinker, gay or straight, as could be found in 1948, largely viewed homosexuality as a handicap to be overcome, and his organization would be a haven for people who struggled with an illness, a force more powerful than its victims, whose first task was to admit they were powerless before it.[7]

Hay's prospectus for Bachelors Anonymous reads in part:

> We, the Androgynes of the world, have formed this responsible corporate body to demonstrate by our efforts that our physiological and psychological handicaps need be no deterrent in integrating 10% of the world's population towards the constructive social progress of mankind. . . .
>
> We aim to aid in the dispelling of this myth [that gays are degenerates because of their physiological and psychological deviations] by attempting to regulate the social conduct of our minority in such matters as, for example, exhibitionism, indiscriminate profligacy, violations of public decency; we aim to explore and promote a socially healthy approach to the ethical values of a constructed pairing between Androgynes; we aim to tackle the question of profligacy and Satyriasis as emotional diseases to be treated clinically.[8]

Membership in Bachelors Anonymous would be secret and members would be protected with fictitious names. If they wished, they would have their real identities unknown even to brother members. Activities would also be secret. Hay's background as a Communist no doubt influenced his conceiving Bachelors Anonymous as a secret society, but how many gays, radical or moderate, could be found who would publicly affiliate with a homophile society?

When Hay founded the Mattachine Society in 1950, it, too, was secret. At a discussion group in 1951, Hay was still struggling with the pathology, not of homosexuality, but of internalized self-hatred, as he struggled with an analysis of the homosexual's position in society:

Homosexuals do not understand themselves and thus it is not surprising that heterosexuals do not understand them either. Because of pent-up frustrations and resentments the Homosexual psychologically rebels and becomes catty and takes on other characteristics of instability. Each person considers himself terribly maladjusted and peculiar. There is now no positive body of information from well-adjusted people. Rather we find case histories of psychopathic and extreme cases that are negative and retrogressive.[9]

Hay would persevere in the face of his own self-loathing, which he would conquer, and in that of the culture's, which proved to be another matter. As early as 1948, he'd heard ominous warnings of what awaited gay men and women as the 1950s dawned. From a friend, he learned of frightening doings in the State Department: Men were being fired because they were gay. The bright dawn seemingly promised in the Kinsey Report might not come after all.

In the days of *Camino Real*, in which an openly gay man would be portrayed for the first time on Broadway without ambiguity, *there was no such thing as gay consciousness* from which an activist or a playwright could draw strength or inspiration. At first, Harry Hay was virtually alone in thinking in terms of a gay community. "[Gays] were the one group of disenfranchised people who did not even know they were a group because they had never formed as a group," Hay said. There was no "gay community" to whom they could look for support, political, moral, or otherwise. When Williams created a gay character for a commercial Broadway play in 1953, he would do so almost in isolation.[10]

Despite—and in some sense, because of—the Kinsey Report, social conditions for gays and lesbians in the late 1940s and early 1950s got worse rather than better. A backlash to Kinsey developed. After all, he had written that in a more tolerant society, homosexual activity would be even more prevalent than his shocking report suggested. If, under the punitive conditions that existed in post-war America, homosexuality was more widespread than had been imagined, then clearly, to the nation's conservatives, homosexuality posed a greater threat to the nation than anyone had thought. J. Edgar Hoover, who had his own sexual secrets, told *Reader's Digest* that Kinsey's work was a threat to "our way of life," and the FBI began assembling a dossier on him and on the Kinsey Institute.[11]

The emerging urban gay and lesbian subcultures that were forming after the war also seemed to pose a threat to the nation, so much so that the armed forces continued the anti-homosexual work they had begun in 1941. During the late 1940s, military discharges for homosexuality numbered about 1,000 per month; by the early 1950s, the number was up to 2,000 (and it would rise again by another 50 percent by the beginning of the 1960s). Between 1947 and 1950, the rate of discharges for homosexuals more than tripled the wartime rate. The armed forces' policy of incarcerating homosexuals either in prisons or in mental wards also carried over into the peacetime civilian world. In October 1950, Donald Vining recorded in his diary a visit to Brooklyn State Hospital near Sheepshead Bay. A friend who worked there escorted him, and as the employee toured Vining through the "violent ward," he told the playwright that, "90 per cent of the cases were homosexual conflicts." Vining was surprised to learn that 60 percent were "cured." How they were "cured," Vining's friend apparently didn't relate.[12]

II

For gay men and lesbians, the years following the war became a Tale of Two Reports: First, Kinsey's, and then, in December 1950, "Employment of Homosexuals and Other Sex Perverts in Government," a report issued by the Senate Committee on Expenditures in the Executive Department. What one document delivered in hope and enlightenment, the other took away.

The Senate investigation had its roots in testimony offered the preceding February by John Peurifoy, an Undersecretary of State who told the Senate Appropriations Committee that of 91 employees fired for "moral turpitude," most were homosexual. Soon after, the Republican National Chairman, Guy Gabrielson, informed by letter 7,000 Republican party workers that "Sexual perverts . . . have infiltrated our Government in recent years . . . [and that they were] perhaps as dangerous as the actual Communists."[13]

Suddenly, dismissals of civilian homosexual employees of the federal executive branch, which had numbered about five per month between 1947 and April 1950 shot up to 60. In May, the Committee heard testimony from Lieutenant Roy Blick, a Washington, D.C., police vice officer, who claimed that 3,500 "perverts" were employed in government agencies (300 to 400 in the State Department alone), and that Washington, D.C., was home to between 4,000 and 5,000 gay men. On June 15, the Senate ordered the Committee on Expenditures in the Executive Department to mount an investigation of perversion in the federal government.[14]

Quickly, the surveillance of law-abiding civilian gay men and lesbians became institutionalized. In order to gather information on the activities of homosexuals, the FBI established contacts with police departments across the country, justifying its actions on the grounds that it was charged with providing the Civil Service Commission with background material on job applicants. Then, the Bureau went further. In an effort to contain and prevent the spread of homosexuality, its regional offices obtained from local vice officers the records of those arrested on morals charges, and forwarded them to Washington—regardless of whether or not the individual had been convicted. The regional offices also collected information on local gathering places of gay men and women and forwarded that information, as well. Merely knowing a homosexual could subject a person to an FBI investigation.[15]

Local police were only too happy to comply with the FBI's request for information and stepped up harassment and arrests. In Washington, D.C., arrests soon exceeded 1,000 a year, and in Philadelphia, they averaged 100 per month. New Orleans, Miami, Memphis, Seattle, Dallas, and Wichita were among the cities where gay men and women became routine victims of police violence and harassment. By the middle of the decade, according to a survey of homosexuals done by Kinsey's Institute, 20 percent of respondents had experienced run-ins with the police.[16]

On December 15, 1950, the subcommittee issued an interim report, "Employment of Homosexuals and Other Sex Perverts in Government." The Introduction concludes with a declaration of the subcommittee's purpose: " . . . sex perverts, like all other persons who by their overt acts violate moral codes and laws and the accepted standards of conduct, must be treated as transgressors and dealt with accordingly."[17]

The report was predicated on the belief that homosexuals were not suitable for employment in any capacity by the federal government, because they were "generally unsuitable," and their "lack of emotional stability . . . and the weakness of their moral fiber," made them serious security risks. Because homosexuals "seldom refuse to talk about themselves," they were especially vulnerable to the interrogation skills of foreign agents. Also, because homosexuals tended to congregate in common nightspots, restaurants, and other gathering places, agents of hostile governments could easily recruit and organize them into espionage rings. Homosexuals were so dangerous, the report stated, that even one could undermine and destroy a government office. A single homosexual employee of the federal government, the report warned, has "a corrosive influence upon his fellow employees. These perverts will frequently attempt to entice normal individuals to engage in perverted practices. This is particularly true in the case of young and impressionable people. . . . One homosexual can pollute a Government office."[18]

Therefore, the report concludes, "homosexuals and other sex perverts are not proper persons to be employed in Government." But due to the laxity of various heads of government offices and departments, too many perverts had been allowed to retain their jobs. However, the subcommittee reports, substantial progress had been made to rectify that dangerous situation. In April 1950, for example, the FBI turned over to the Civil Service Commission all the police records of people charged with sex offenses in the District of Columbia. The subcommittee also found that the legislative branch had been as lax in its vigilance as the executive had but that, " . . . since the initiation of this investigation all known perverts in the legislative agencies have either been removed or the cases are being given active consideration."[19]

Furthermore, the subcommittee recommended beefing up the Criminal Code as it pertained to Washington, D.C., increasing the penalties for "lewd, obscene or indecent" acts and transferring responsibility for prosecuting these offenses in the District to the office of the U.S. Attorney. At the encouragement of the subcommittee, the District's Superintendent of Police had enlarged the detective squad that was assigned to gay "vice" and promised to add still more manpower once men were properly trained in the art of entrapment.[20]

The subcommittee concluded on a dual note of satisfaction and warning. First, it was pleased to report that "considerable progress is being

made in removing homosexuals . . . from positions in the government."
It pointed to the fact that "action has been taken in 382 sex perversion
cases involving civilian employees of Government in the past 7 months,
whereas action was taken in only 192 similar cases in the previous 3-year
period from January 1, 1947 to April 1, 1950." In its final paragraph,
however, the report warned that responsible government officials must
continue to "maintain a realistic and vigorous attitude toward the problem of sex perverts in the Government. To pussyfoot or to take half meas-
ures will allow some known perverts to remain in Government." Finally,
it promised that the subcommittee would reexamine the problem "from
time to time," to ensure that the public interest was being served, and
that the nation was being kept secure from the threat of homosexuals at
work anywhere in the federal government, from its lowest ranks to its
highest offices.[21]

Instances of homophobia in the federal government, amplified by the
subcommittee's report, continued to grow in 1951. In March, the
State Department dismissed a vice consul and three other members of
the U.S. consulate in Hong Kong for homosexuality, and the story
was reported on the front page of *The New York Times*. At the end of
April, the paper reported a statement from J. Edgar Hoover to the effect that since the beginning of the month, the FBI had identified
406 homosexuals in government service. In October, four State Department employees in Korea resigned after having been accused of
"homosexual acts," according to the *Times*. In December, it reported
that at least four government agencies were subjecting employees to
lie detector tests to determine if they had ever been arrested, associated with Communists, or were homosexuals. By April 13, the roster
of State Department employees who had been fired from their jobs
either for being homosexual or accused of being, had reached 425
since 1947, and from 1947 through mid-1950, 1,700 job applicants
were denied work in the federal government because they were homosexual. On June 26, 1952, the *Times* reported that in 1951, the State
Department discharged two employees on charges of homosexuality,

and that when confronted with similar allegations, another 117 had resigned.[22]

The shrillness and paranoia of the Senate report also found its way into the popular press. In 1951, Crown published *Washington Confidential* by Jack Lait and Lee Mortimer, which purported to expose the seamy sexual underside to the capital. The authors wrote that the city was in danger of being overrun by perverts:

> . . . the chief meeting place is in leafy Lafayette Square, across Pennsylva-nia Avenue from the White House. They make love under the equestrian statue of rugged Andrew Jackson, who must be whirling on his heavenly horse every time he sees what is going on around his monument. . . . How the fairies happened to pick this place for their rendezvous, and how the cops let them get away with it, no one can trace. . . . Many rich fairies and lesbians live in expensive remodeled Georgetown homes, the nearest thing to a Left Bank neighborhood. This is also a left-wing center. . . . Fags like a restaurant known as Mickey's, behind the Mayflower. . . . One night two Congressmen, a couple of army officers and two young servicemen were mixing beer and gin there, and kissing each other. They also swish around the Sand Bar in Thomas Circle."[23]

Meanwhile, the Broadway theatre was playing its part in the demo-nization of homosexuals. It is important to note that producers of the plays of these years that featured homosexual characters did not set out to demonize anybody; indeed, many of the producers' best friends probably were homosexuals. But in choosing the plays they chose, commercial producers replicated the prevailing notions that homosexuals were dan-gerous and that to be unfairly accused of homosexuality was a disaster that could have deadly results.[24]

The first depiction of gay men on Broadway in the 1950s was on the face of it rather harmless. *Season in the Sun* opened in late September 1950. The characters included two gay men variously described by critics as "two men who should have been women," "two amusing effeminates," and "two . . . queens (a type which flourishes on Fire Island like Eel-grass." The play was written by the *New Yorker* drama critic Wolcott Gibbs. His description of the gay pair reads, "They would have no trou-ble at all flying in and out of windows."[25]

The first Broadway play of the 1950s that depicted homosexuality as a problem was a 1951 revival of Mordaunt Shairp's *The Green Bay Tree,* which, in its initial Broadway production of 1933, introduced Laurence Olivier to American audiences. The play's leading character, Mr. Dulcimer, is an independently wealthy, effete Englishman who Shairp describes this way: "He has a habit of looking at you from under his eyes, and though a complete dilettante, he has an alert, vibrating personality. A man who could fascinate, repel and alarm."[26] Several years prior to the play's first scene Dulcimer literally purchased a boy from his drunkard, working-class father. He is raising Julian, now a young man, to be like him: to master the fine arts of classical music, floral arrangements, needlepoint, and expensive dining. Julian, however, falls in love with a young woman, Lenora, and vows to leave the sybaritic life, get a job, and marry. Dulcimer, however, convinces him to stay (an easy task, as it turns out, for Julian is not used to such quotidian details of the workaday world as having a job). In despair for his son, Julian's natural father appears and shoots Dulcimer with the villain's own jeweled revolver. But Dulcimer's repellent charms have worked well. In the final scene, we find Julian in Dulcimer's Mayfair flat, refusing Lenora's insistence that he give up the fortune that "Dulcie" left him. As Lenora leaves him for the last time, Julian begins arranging irises in an amber vase.

Although homosexuality is never explicitly mentioned or shown in *The Green Bay Tree,* the play's message is transparent: Homosexuality is an evil, and a dangerously communicable one (the title is a line from Psalm 37, "I myself have seen the wicked in great power and spreading himself like a green bay tree"). In the original New York production, the subtext was clear enough for the critics to hail the play's cautionary note. In *The New Yorker,* Wolcott Gibbs had written,

> The subject of abnormality is usually discussed on stage with such a misty delicacy that it is hard to tell just exactly what the author had on his mind, or else with a humor hearty and explicit enough to engage the indignant attention of the Police Department. It is a relief to see it treated for once, in *The Green Bay Tree,* at the Cort, without either coyness or the nasty joviality of a Minsky blackout.[27]

The gay theatregoers of New York certainly knew what the play was about; on more than one occasion they filled the balcony.[28]

Unlike the 1933 production, the 1951 revival attempted to remove even the subtextual meaning—leaving the critics, especially those who'd seen the first production, baffled—not by the play's theme, but by the producer's attempt to cover it up. For those who didn't see the original, John Chapman, writing in the *Daily News,* was happy to fill them in: "People who were playgoing in 1933 will remember that 'The Green Bay Tree' is about a wealthy, intelligent old pervert and his adopted son. The young man obeys the natural urge to fall in love with a girl, and the drama concerns the older man's methods of destroying this love." In *The New York Post,* Richard Watts, Jr., also spelled it out in a bemused tone: "What [theatergoers last night] saw was a rather tedious drama . . . about the struggle for the soul of a dull young man between a debonair hedonist who wanted him to lead a life of luxury, and a girl, who wanted to make a veterinary out of him. I assure you it was at one time more than that. For 'The Green Bay Tree' was a fine play, a chilling and sinister drama about a ruthless, exquisite homosexual who dominated the life of a well-meaning but weak young man . . ."[29]

The result was a reticent rendition of a play about an evil homosexual that was outed by the critics. This production of *The Green Bay Tree,* on the heels of the Senate report, seemed to want it both ways: a play about how one homosexual can corrupt an innocent youth while trying hard not to admit what Dulcimer's vice was (a difficult task, given lines such as Dulcimer's, "[Mother] was the only woman who ever meant a thing to me."). As a part of mainstream culture, Broadway was reinforcing Cold War attitudes toward homosexuals.[30]

The pattern of harassments, arrests, and government-sponsored witch-hunts continued into 1952. That year, the American Psychiatric Association categorized homosexuality as a sociopathic personality disorder in the first edition of its *Diagnostic and Statistical Manual of Mental Disorders (DSM–1),* adapting the terminology and thinking pioneered by armed forces psychiatrists during the war. In the spring, Mattachine distributed a leaflet around Los Angeles titled, "Your Rights in Case of Arrest," that outlined the ways to respond to police arrests and harassment. It depicts quite clearly the dangers that gay men and women faced every day, even for merely congregating in a public place.[31]

1. If an officer tries to arrest you, he should have a warrant unless a misdemeanor (minor violation) or a felony (serious offense) has

been committed in his presence or he has reasonable grounds to
believe the person being arrested is guilty.

2. If he has no warrant ask what the basis of the arrest is. If it is not
   explained as in No. 1 above, go along but under protest made before a witness if possible. DO NOT RESIST PHYSICALLY.

3. GIVE NO INFORMATION! You may, but do not have to, give
   your name and address. Do NOT talk to any policeman.

   > Q: *Why did you commit this crime?*
   > A: "I'm not guilty and I'd like to speak to my attorney,
   >   please."
   > Q: *How long have you been a lewd vagrant?*
   > A: "I'm not guilty and I'd like to see my lawyer before making
   >   a statement."
   > Q: *Have you been arrested for this before?*
   > A: "I'm not guilty and my attorney would rather I speak
   >   through him."
   > Q: *Nice day, isn't it?*
   > A: "I'm sorry, but I'd like a lawyer's advice before making a
   >   statement."

4. Deny all accusatory statements by arresting officers with, "I'm
   not guilty and I'd like to contact a lawyer." Otherwise your silence before witnesses can be construed in court as assent.

5. If an officer insists on taking you to jail, ask when you are
   booked (registered) what the charges are.

6. Insist on using a telephone to contact your lawyer or family.

7. DO NOT SIGN ANYTHING. Take numbers of arresting officers.

8. You have a right to be released on bail for most offenses. Have
   your attorney make the arrangements. Or you can ask for a bail
   bond broker. For a fee, he will post (deposit with the police) the
   amount needed for your release.

9. Under no circumstances have the police a right to manhandle,
   beat or terrorize you. REPORT ALL SUCH INCIDENTS.

10. If you do not have an attorney by the time you are required to
    plead guilty or not guilty, remember this:

    a. You are entitled to a copy of the charges made against you.
    b. You are entitled to have a lawyer. Ask for a postponement
       until you get legal representation.

11. PLEAD NOT GUILTY.

12. Ask for a trial by jury unless your lawyer advises otherwise.

13. You are not required to testify against yourself in any trial or
    hearing.

14. If you are questioned by a member of the FBI, you are not required to answer. Immediately consult an attorney so that your
rights may be adequately protected.[32]

New York City, where rehearsals for *Camino Real* would shortly begin,
was no safer haven for gay men or women than Los Angeles. Members of
the vice squad hid in subway men's rooms, waiting to witness or entrap
men engaging in consensual sex acts; in the men's room in Grand Central
Station, they watched from behind a two-way mirror. Even the bars
could be dangerous meeting places. It was illegal for gays to congregate in
bars and for bar owners to allow it, and plainclothes policemen would
lounge against a table or wall, waiting for a chance to make an arrest. At
first, the cops were almost amusingly easy to spot, as they often wore un-
fashionable black shoes, tended toward chubbiness and nursed a single
drink for hours. In time, however, younger lawmen dressed in jeans and
leather jackets replaced their hapless predecessors. Rather than waiting to
be approached, these more aggressive cops enticed men into entrapment.
Building inspectors closed gay bars for the smallest infractions, including
cigarette butts on the floor, and fire inspectors closed bars when they decided the establishment was overcrowded. By the time the World's Fair
was held in Flushing Meadows, every solely gay establishment in Man-
hattan had been closed, with the exception of the Everard Baths.[33]

Bars that were not gay but that were nonetheless popular homosexual
gathering places took to discouraging gay men's patronage. The Oak
Room at the Plaza Hotel no longer allowed men to stand at the bar if
they were unescorted by women. Another popular place, P.J. Clarke's, did
the same. Gay men and women were no safer in the summer communities of the Pines and Cherry Grove on Fire Island. On crowded weekend
nights, boats of Suffolk County policemen would ferry over from the
mainland. Literally beating the bushes with billy clubs, they bashed men
having sex and dragged them off to jail. Many of the men's names would
appear in the next day's newspapers—not as victims of police brutality,
but as perpetrators of sex crimes.[34]

1952 saw the Broadway revival of another hit from the late 1930s, Lillian Hellman's *The Children's Hour*. Directed by Hellman and produced
by Kermit Bloomgarten, the revival was offered as a liberal's retort to the
McCarthyite tactics of innuendo and guilt by association. The play takes

place at the Wright-Dobie school for girls, where a pathologically nasty young student spreads rumors of a lesbian relationship between the school's two proprietors. We are meant to be outraged by the girl's viciousness (after all, what could one say about a person worse than that she was a lesbian?) and by the community's willingness to believe the slander without proof. In 1952, this was bad news enough for lesbians. But Hellman muddied the waters further in the last scene, when Martha Dobie admits to Karen Wright that the rumors are true, at least as far as she's concerned:

> Martha: *I have loved you the way they said.*
> Karen: You're crazy.
> Martha: There's always been something wrong. Always—as long as I can remember. But I never knew it until all this happened. . . . You're afraid of hearing it; I'm more afraid than you. . . . You've got to know it. I can't keep it any longer. I've got to tell you how guilty I am.[35]

Martha exits. A moment later, the fatal pistol shot rings out.

Hellman's purpose was not to point out the evils of lesbianism; it isn't a "problem play" in that regard. Hellman's problem is the purposeful spreading of vicious lies. But in 1952, the famously liberal Hellman, who refused to name names before the House Committee on Un-American Activities, and her equally liberal, well-meaning producer, found nothing wrong in using lesbianism as an example of heinous behavior. The reviewer for *Women's Wear Daily* understood:

> Gossip about sexual aberrations, actual and implied, has become more prevalent in our depraved plays of 1952 than it ever was in 1934. The implied Lesbianism between the teachers and the nasty gossip-mongering by the patrician grandmother of the malicious little girl seem completely topical and valid today.[36]

The federally sponsored witch-hunt of gay Americans reached its height on April 27, 1953, when President Eisenhower issued Executive Order 10450. It instructed heads of departments and all those who did hiring, that "sexual perversion" was not only sufficient, but necessary grounds for disqualification from federal jobs. In the next 16 months, homosexuals were removed from their jobs at the rate of at least 40 per month. The effect of Eisenhower's Executive Order, however, did not stop with federal employment. Every private company with which the federal government had a contract had to comply. By the middle of the decade, the policy had trickled down to state and local governments, extending the ban on homosexuals' right to work to over 12 million workers—more than 20 percent of the national workforce. Moreover, these workers were now often required to sign "moral purity oaths," in order to get or keep jobs. Organizations such as the American Red Cross adopted similar policies.[37]

Late in 1953, a new playwright emerged with a hit that would run for 712 performances. Robert Anderson's *Tea and Sympathy* concerns a young man at yet another private school in New England, Tom Lee. He likes classical music, is not competitive at sports, and has a sissy walk. Needless to say, he is tormented by his peers. He's tried everything to prove to himself that he's normal, but an anxiety-ridden tryst with a local prostitute ends in humiliating failure. His housemaster, a hale fellow who spends more time mountain-climbing with the boys than he does with his wife, is particularly cruel to him. Only the housemaster's wife, Laura, takes pity on him. "I wish they'd let me kill myself," Tom tells her in despair, and, taking matters into her own hands, Laura leads him to her bedroom in the closing moments. She will teach him how to make love to a woman and, in her famous last line, implores him, "Years from now—when you talk about this—and you will—be kind." All, we assume, will turn out well, and Tom will go on to marry and produce a crop of more tolerant children. Perhaps he will be elected to the Senate.[38]

The misunderstood, passive Tom must share the hero's laurels with Laura, who senses from the beginning that he's a normal, red-blooded American boy, just more sensitive than most. The villain is the housemaster, Bill, who not only taunts the boy, but is also cruel and distant to his wife. Passion and gentleness both have long been absent from their marriage; it seems that Bill can't bring himself to touch her. "Now it's long separations," she laments to him, "and then this almost brutal coming together . . ." (55).

Bill's chief failing, however, is not his cruelty. By Act Three, the wise Laura understands. "Did it ever occur to you," she says to Bill, "that you persecute in Tom, that boy up there, you persecute in him the thing you fear in yourself?" The stage direction reads, *Bill looks at her for a long moment of hatred. She has hit close to the truth he has never let himself be conscious of. There is a moment when he might hurt her, but then he draws away, staring at her.* "This was the weakness you cried out for me to save you from, wasn't it," Laura says to him (84). Their marriage over, the villain can only twirl his mustache impotently and exit. The sensitive young boy is cleared of all charges of homosexuality, while the "problem," that is, the homosexual, is exposed. He will no longer blight his wife's life, and no doubt he will leave the school, as well, and will cease to pose even a latent threat to the schoolboys.

This was a play that America could embrace, and it did. The critics greeted *Tea and Sympathy* ecstatically. "Last night's new entry, 'Tea and Sympathy,' can most simply and briefly be described as a triumph," William Hawkins wrote in the *New York World-Telegram.* The "beautifully written" play, John Chapman declared, has a theme of "homosexuality—or the suspicion thereof; and such a play, if it is not to be merely tawdry and cheaply sensational, must be made with good taste and great adroitness . . . [Anderson] has written like a veteran and like a poet . . ." Richard Watts described the play as "a sensitive and effective sentimental drama of unquestionable box-office magnetism." These were the kind of reviews that even Tennessee Williams seldom got.[39]

As simple as it is to make light of *Tea and Sympathy,* the play had a message that was useful for America to hear in the fall of 1953. "Manliness is not all swagger and swearing and mountain climbing," Laura tells Bill before exposing his self-loathing. "Manliness is also tenderness, gentleness, consideration" (83). This was not the moral, however, that the public took to its heart. The boy wasn't a queer, and that was what mattered.

▬▬▬▬▬▬

The persecution of homosexuals in the late 1940s through the 1950s was not occurring in a vacuum. By 1946, a larger cultural reaction set in in response to the vast changes in gender roles that occurred during the war. Millions of women who had found freedom and fulfillment in work outside the home during the war lost their occupations. It was more important, the culture declared, that their husbands be given back their jobs. Women began seeing reminders, in newspapers and magazines, that their proper sphere was the kitchen and nursery. News articles began linking juvenile delinquency with the absence of mothers from the home, and mothers were treated to photographs of children smoking cigarettes, as a warning of what would happen if women neglected their domestic duties. Magazines that during the war promoted products that allowed a woman to do the housework quickly, such as canned soup, began printing recipes that took an entire day to prepare.[40]

Not only were women and homosexuals being put in their place (the kitchen, bedroom and nursery, and the invisible world of non-being), whole notions of what aspirations were appropriate and which behaviors acceptable were being set in place, notions that constricted the dreams of the New Deal and even negated some of the avowed purposes of the war. The suppression of these dreams and aspirations is in large part the subject of *Camino Real.*

III

Camino Real opened at the Martin Beck Theatre on March 19, 1953. Williams viewed the play as something of a breakthrough, an example of the "plastic theatre" he had advocated in his production notes to *The Glass Menagerie.* He filled it with the color, movement, music, and theatricality he believed were the theatre's strong suit. Unlike *The Glass Menagerie, Camino Real* could not mistaken for typical American realism. Williams wrote in *The New York Times* four days before the play's Broadway premiere,

> More than any other work that I have done, this play has seemed to me like the construction of another world, a separate existence. [. . .] To me, the appeal of this work is its unusual degree of freedom. When it began to get underway I felt a new sensation of release, as if I could "ride out" like a tenor sax taking the breaks in a Dixieland combo or a piano in a bop session. [. . .] My desire was to give these audiences my own sense of something wild and unrestricted that ran like water in the mountains, or clouds changing shape in a gale, or the continually dissolving and transforming images of a dream.[41]

The play, arranged in 16 "blocks," or scenes, may be the dream of Don Quixote who, with the faithful Sancho Panza, opens the play (in the revised version published in *The Theatre of Tennessee Williams*) by striding down the theatre's center aisle, dressed as an old "desert rat." The pair has come up from the desert that surrounds the Camino Real. Sancho disappears, frightened by his map's description of the street: "Halt there [. . .] and turn back, Traveler," he reads, "for the spring of humanity has gone dry in this place and . . . there are no birds in the country except wild birds that are tamed and kept in—*Cages!*"[42]

Quixote, however, is unperturbed, and settles down for the night. "And my dream will be a pageant," he says, "a masque in which old meanings will be remembered and possibly new ones discovered [. . .]" (437).

The old meanings take the form of characters from Romantic literature, or iconic figures of that era: Jacques Casanova, Marguerite Gautier, George Gordon Lord Byron. They stay on "the luxury side" of the Camino Real, at The Siete Mares, an elegant hotel with a terrace on which they eat, drink, and passively await escape on an airplane called The Fugitivo, which makes unscheduled arrivals and departures.

The other side of the Camino Real is Skid Row, inhabited by the poor and dispossessed, by the beggars, gypsies, and assorted riffraff who are scrupulously barred from the Siete Mares by the large figure of Gutman, the strongman in a linen suit who presides over both sides of the street. Skid Row is home to the various services that cater to its poor denizens: a fleabag hotel, a gypsy's stall, and a pawnshop whose neon sign glows in pink, green, and blue pastels, "Magic Tricks Jokes." There is also a fountain on Skid Row, but it has gone dry. The only fountain that still gives forth water is in the Siete Mares Hotel, where it is patrolled around the clock by armed guards. The street is surrounded by a high wall near the top of which is an archway that leads out of the town to the desert and snow-capped mountains beyond, referred to in the play as "Terra Incognita."

Often, the first thing that happens in a play is thematically significant; in *Camino Real,* after the brief Don Quixote prelude, the first action that we see is a meeting between the eighteenth-century lover and writer, Jacques Casanova, and Prudence Duvernoy, the old friend of Marguerite

Gautier. Although he is stopping at the exclusive Siete Mares, where he is courting Marguerite, Casanova is living in reduced circumstances. In order to pay his bills at the hotel, he attempts to hock his silver Boucheron snuffbox at the Loan Shark's, but is offered an insultingly low price. He meets Prudence at the dry fountain on Skid Row, where she is searching for her lost poodle. She wants him to give Marguerite a message from a former customer: The wealthy old man wants her back on any terms.

Casanova is appalled, but Prudence will not be put off. Marguerite has left the wealthy old man for a young one with no fortune simply because she thinks she loves him, and, Prudence insists to Casanova, one can't do that anymore, you've got to be realistic on the Camino Real:

> Oh, I've told her and told her not to live in a dream! A dream is nothing to live in, why, it's gone like a—[. . .] Times and conditions have undergone certain changes since we were friends in Paris, and now we dismiss young lovers with skins of silk and eyes like a child's first prayer, we put them away as lightly as we put away white gloves meant only for summer, and pick up a pair of black ones, suitable for winter . . . (442)

What is Casanova's response? Does he mount a ringing defense of the woman he loves? Does he issue a denunciation of Prudence's accomodationist pragmatism? No. He flees. He, too, is living in a dream: His credit in the Siete Mares restaurant and bar has been cut off, and he is dependent on Marguerite to pay his bills. He has been expecting for the longest time (he says) a letter with a check that will clear all his debts. Gutman has warned the great lover that if the letter doesn't arrive tonight, Casanova will have to transfer his patronage to the fleabag Ritz Men Only, on Skid Row.

In the face of dual provocations—Prudence's and Gutman's—Casanova can only, in the first instance, run away, and in the second, sink, trembling, into a chair. Casanova—adventurer, gambler, libertine, writer—is reduced to flight, fear, and inaction. What is Williams saying about these Romantic figures?

Next, a delirious young man called The Survivor, his clothes reduced to rags, his skin burned black by the sun, stumbles into the plaza. He alone has returned from an expedition of young people into the Terra

they represent are a sad and passive lot. Except for one: the Baron de Charlus.

Although he's an aristocrat, the Baron does not stop at the Siete Mares. Unlike the other Romantics, he prefers the life on Skid Row. If the Siete Mares, with its stone facade and glass-topped white wrought iron-tables is where the effete and impotent stay, Skid Row, according to Williams's set description, *"should have all the color and animation that are permitted by the resources of the production"* (462). And if the Baron, whom Williams describes as *"an elderly foppish sybarite in a light silk suit"* (464), is dedicated to little else than pleasure, at least he is more democratic in his search for it than are his Romantic colleagues, who either enjoy their gilded cage, or are too afraid to leave it.

The Baron is on the prowl for sex. When A. Ratt, the proprietor of the Ritz Men Only, asks the Baron why he comes to the seamy side of town rather than have his "joy rides" at the Siete Mares, the Baron is straightforward:

> They don't have Ingreso Libero at the Siete Mares. Oh, I don't like places in the haute saison, the alta staggione, and yet if you go between the fashionable seasons, it's too hot or too damp or too appallingly overrun by all the wrong sort of people who rap on the wall if canaries sing in your bedsprings after midnight. (465)

It seems that the Baron is about to pick up "a wild-looking young man of startling beauty" from the wrong side of the tracks called Lobo, who is trailing him. "Is he attractive?" the Baron asks A. Ratt, who replies, "That depends on who's driving the bicycle, Dad" (465). But first, the Baron makes the acquaintance of Kilroy, the innocent, All-American former Golden Gloves boxer with an enlarged heart of gold. Kilroy inquires after the local hot-spots; in his reply, the Baron transports us directly to New York City, 1953, as he describes the gay bars on the East Side:

> Oh, the hot spots, ho ho! There's the Pink Flamingo, the Yellow Pelican, the Blue Heron, and the Prothonotary Warbler! They call it the Bird Circuit. But I don't care for such places. They stand three deep at the bar and look in the mirror and what they see is depressing. One sailor comes in— they faint! My own choice of resorts is the Bucket of Blood downstairs from the "Ritz Men Only." (469–70)

The Baron gives Kilroy the once-over, and the young man's innocence appeals to him—but only so far. Lighting a match, he looks into Kilroy's eyes: "The eyes are the windows of the soul, and yours are too gentle for someone who has as much as I have to atone for" (470). With that, that Baron bids the young man au revoir, and exits. Moments later, he is killed by the Streetcleaners, Williams's metaphor for the violence visited on the persecuted of the world. So much for the first openly gay character in a Broadway play who clearly and guiltlessly proclaims his sexual allegiance.

The Baron's appearance, brief as it is in *Camino Real,* is a landmark in Williams's work and in the history of gay characters in the American theatre. At the height of the McCarthy era, three weeks before President Eisenhower issued Executive Order 10450, Williams presents not only an openly gay character, but one with the exotic sexual tastes of a sadomasochist. Williams could not have been more open, or less judgmental, about what the Baron is looking for when speaking to A. Ratt: "You know the requirements. An iron bed with no mattress and a considerable length of stout knotted rope. No! Chains this evening, metal chains. I've been very bad, I have a lot to atone for . . ." (464–5).

Yet, most gay critics who have written about Williams since the 1970s ignore the Baron. Because he is borrowed from Proust's *la Recherche du Temps Perdu,* the Baron is, according to the critic Georges-Michel Sarotte, "disembodied," that is to say, not really a character, and not really there. He is

> a soul in torment surrounded by a halo of unreality and surrealism. Like Blanche's husband, Charlus is not really *physically* present on stage because he is part of the literary history of another country, and because he is in hell. . . . Charlus is a grotesque.[44]

Sarotte's belief in the Baron's non-being is a good demonstration of the trouble literary critics have understanding the theatre, and why most theatre practitioners tend to have little use or patience for their work. While he's being fitted for his pale yellow suit and given a small hand mirror with which to inspect Lobo, it would be difficult to explain to an actor playing Charlus, how it is that he's not physically present onstage. To the extent that Charlus does exist, Sarotte writes, he is, at best, a pas-

sive victim. Worse, Sarotte says, through the Baron, Williams "links sexuality, and particularly homosexuality, with expiation."[45] Sarotte, at least, remembers the Baron and *Camino Real;* Nicholas de Jongh, in *Not in Front of the Audience: Homosexuality on Stage,* never mentions him, rendering him totally invisible, although he writes about Williams at length.
David Savran is unequivocal:

[Williams's] homosexuality is the site of manifold contradictions, articulated by the unstable and fluid difference between secrecy and disclosure, between his ability to write about his sexual desire to his gay friend [Donald Windham] and his inability to speak about it openly . . . for many years, to the theatregoing public. . . . Instead, Williams's homosexuality is endlessly *refracted* in his work: translated, reflected, and transposed.[46]

In one sense, Savran is correct: Williams doesn't "speak about" "homosexuality" in *Camino Real;* he creates a gay character. This is what playwrights do while critics "speak about homosexuality." Savran goes further in describing what he views as Williams's homophobia. "[The] habit of constructing a plot upon a 'guilty secret' that is never entirely divulged certainly encodes Williams's own 'guilty secret' and the impossibility of its revelation during the 1940s and 1950s as anything other than an 'ugly truth.'" What is ugly in *Camino Real* vis-à-vis homosexuality is neither the Baron nor his sexual tastes, but his murder by the state. The fact that critics who denounce "invisibility" in Williams render the Baron invisible isn't too pretty, either.[47]
Robert J. Corber, in *Homosexuality in Cold War America: Resistance and the Crisis of Masculinity,* finds that, " . . . none of Williams's plays from the fifties concerns unequivocally gay male characters. . . ." Nor does John Clum mention the Baron in any of his writings on Williams—except by omission, when he writes that Williams didn't decide to "move his openly gay characters from the exposition to the stage" until "safely after Stonewall." Clum writes, "Williams was compelled to write about homosexuality, but equally impelled to rely on the language of indirection and heterosexist discourse. Gaining the acceptance of that broad audience meant denying a crucial aspect of himself." As far as the Baron de Charlus is concerned, however, Williams not only denied nothing, he defied the 1953 mainstream audience to deny a powerful

and, to many, unappealing kind of sexual urge. The Baron is no closeted Brick Pollitt. He is nothing other than what he is; he does not apologize, he does not hide.[48]

─────────────

In Scene Four of *The Glass Menagerie*, Tom also talks of canaries: They were transformed from goldfish and fly away to freedom. That is what the Baron desires in sex: freedom to be who he is, free from the judgment of the "squares" against whom Williams so often revolted. The critics who condemn the Baron and find his sexual tastes "non-affirmative" or ignore him altogether ally themselves not with post-Stonewall openness and acceptance but with those who make negative moral judgments on exotic sexual tastes: that is, with the ideology of the broad audience to whom Williams is accused of pandering.

Why do these critics ignore and dislike the Baron so? Perhaps it has something to do with the gritty nature of his desire. They may find his description of the Bird Circuit distasteful. And while some critics may see in the Baron's violent end a recapitulation of the "homosexual problem play," in which the problem is solved by the death of the homosexual, Williams could reply that the death of the Baron is not expiation, but an accurate account of the dangers experienced by openly gay men in the 1950s who risked arrest merely for standing on a street-corner or sitting in a bar. If the critics are either squeamish or so intent on political correctness that they indulge in a puritanism far more intense than Williams's own, this is not the playwright's problem. They view the Baron in isolation, and neglect the actions of *Camino Real*'s other Romantic figures compared to his.

Of the play's Romantic figures, the Baron is the one democrat. He freely consorts with the "riffraff" on Skid Row, while Casanova and Marguerite acknowledge neither their existence nor their suffering. Indeed, it may be the Baron's own penchant for suffering that makes him aware of the desire, suffering, and existence of those who live on the wrong side of the tracks.

The Baron speaks easily and openly with Kilroy, one man to another. Casanova, fearful of Gutman and the Streetcleaners, addresses the young man only in a whisper while on the lookout for armed guards. Casanova speaks to Kilroy of the importance of romance, of their both being "Travelers born," satisfied by nothing, always hopeful. "The exchange of serious questions and ideas, especially between persons from opposite sides of the plaza, is regarded unfavorably here," he confides to Kilroy, while never daring to exchange so much as a glance (472). Casanova, in the end, espouses a Romanticism that, unlike the Baron's, fears being seen in the streets. He is a Romantic afraid of authority, terrified of the Terra Incognita beyond the Camino's walls, unwilling to speak the forbidden word, *Hermano*. What was once a daring, muscular Romanticism that endeared the real Jacques Casanova to Voltaire, that engaged him in dueling, spying, and composing dangerously satirical pamphlets, is reduced here to hankering for his reflection in the eyes of a woman, which he prefers to think of as love. Casanova is no more than an advertisement for the past. (That Williams knew exactly to what degree he was reducing Casanova is made clear by the fact that, in preparation for writing *Camino Real*, he read all 12 volumes of Casanova's memoirs.[49])

For her part, Marguerite, while no braver than Casanova, can at least acknowledge her situation. Rather than carrying the banner of Bohemia into the enemy camp, as Casanova emptily suggests, she prefers to believe that, these days, "Bohemia has no banner. It survives by discretion" (493).

When the mysterious airplane called The Fugitivo arrives at the Camino Real to take on passengers for an unspecified destination, Marguerite shrieks, begs to be taken on board. She orders Casanova to her room for her jewels, which the Pilot of the craft will not accept; she then sends her accommodating lover to change her francs—also unacceptable—into dollars. When the Pilot demands her missing papers, she turns "savagely" on Casanova, tearing open his coat and seizing his papers in hopes of getting them past the Pilot. After The Fugitivo departs without her, all she can do is lean on her lover and murmur, "Lost! Lost! Lost! Lost!" (524). Is this the brave Marguerite Gautier of *Camille?* Yet it is hard to feel much for the practical old Prudence Duvernoy and her friend Olympe, Marguerite's old companions from her Paris days, when they appear just in time to board The Fugitivo, elbowing Marguerite aside as they go.

There is another Romantic figure on the Camino Real; he falls between the instincts of the Baron on the one hand and of Casanova and Marguerite on the other. At first, it seems as if George Gordon Lord Byron is also intent on fleeing. In Block Eight, he appears with his luggage (consisting mainly of caged birds), and declares his intent to leave the Camino Real. No longer arrogant, proud or even confident, Byron knows he is not longer what he once was: a true Romantic, a Revolutionary. "The luxuries of this place have made me soft," he tells Gutman. "The metal point's gone from my pen, there's nothing left but the feather." His solution? Not to go forward, to explore new territory, to be a Childe Harold crossing the Alps, but to retreat into his past. "From my present self to myself as I used to be!" Nothing could make Gutman happier: "That's the *furthest* departure a man could make!" (503–4). There is nothing less threatening to a totalitarian regime than a nostalgic Romantic.

Yet Williams does not dismiss this Byron as a decadent. For Byron knows what he's been, and what he's become. He sees the way that he's allowed himself to be distracted by the by-products of fame, by the dances, salons, and women who flung themselves at him. He has, he says, ceased listening to his heart, and therefore the time is ripe to strike out from the Camino Real and seek inspiration again at the foot of the Acropolis. He may not hear it there, and certainly a return to the past is not as likely to provide inspiration as a foray into the future might, but he is the only Romantic other than the Baron—dead by now—who can say, "*Make voyages!—Attempt them!*—there's nothing else . . ." (508).

Flight, so often characters' instinctive reaction in a Williams play, is one thing; making voyages is another. Making voyages, as Byron puts it, means striking out for parts unknown with a bravery ready to meet them, with a spirit willing to accommodate the new. Flight, in Williams's dramatic world, means fleeing headlong into the unknown because of fear of the known. No thought is given to flight; it is the opposite of a journey made in good faith. Flight is a frightened reaction to a fearful moment, the lack of resolution in the face of danger, the unthinking instinct toward self-preservation. The instinct is human, so it cannot be condemned. Making voyages, however, is a conscious choice that, like the one to stay and struggle, recognizes the humanity in oneself and in others. In this respect, Gutman is an over-seer in more than just the way he introduces and comments on each scene. He is also the overseer of the meek slaves who live in the Siete Mares. He is the Master of both communities who populate the Camino Real—the poor and dispossessed who live on Skid Row, and the dispossessed of spirit who inhabit the hotel with its guarded fountain, who nightly drink themselves into consolation on other people's money.

Like Byron, the Baron makes voyages. Williams suggests this in the language with which he furnishes A. Ratt, as he tempts Charlus with a vacancy in the Ritz Men Only: "A little white ship to sail the dangerous night in" (465). Byron's voyages, one might argue, are solitary, taking him into the desert wasteland, away from the crises on the Camino Real. The Baron's voyages are interior excursions of self-discovery, of the boundaries of personal limitations, but they also connect him to other people—to the wild young Lobo, to Kilroy. They do not take him away from the Camino Real; they take him more deeply into it. Contrasted with the other Romantics, the Baron is *engaged.* One should be careful not to tread too far, however: the Baron's human connection is forged through the need for human contact more than for the sake of a specific political belief. But it is contact nonetheless, and suggests at least the potential of a social awareness that is absent in the other Romantics, all of whom, in their original incarnations, were creatures of profound political implications.

Camino Real was, as much if not more than other Williams works, the result of a genesis that can be viewed across a long trail of manuscripts. One of these was a one-act version performed at The Actors Studio in 1949, *Ten Blocks on the Camino Real.* The Baron appears in this version, as well. By the time Williams expanded the play to full-length, he expanded the Baron's role, too. In the one-act, there is no Lobo, and the Baron is not explicitly cruising, as he is in *Camino Real.* In the later version, which Williams was writing specifically for Broadway, the play-wright gave the Baron his speeches about The Bucket of Blood, the Bird Circuit, and his requirements for a bed at the Ritz Men Only. He also identifies the Baron specifically as de Charlus. In other words, at the height of McCarthyite paranoia and homophobia, looking forward to a

Broadway production, Williams turned an implicitly gay character into one whose identity was explicit.

Voyages were at least as common to Williams's nature as flight. The Baron's objection to the lack of "Ingreso Libero" at the Siete Mares and the complaints of the hotel's patrons to canaries singing in his bedsprings, is reminiscent of Williams's own relations with hotel clerks and detectives. He, like any sexually active gay man of his era, knew what it was like to be watched by surveillance-minded authorities. Years after *Camino Real,* he wrote of his days in the Shelton Hotel in New York in 1946, where he and a friend would find tricks in the steam room.

> After almost every session in that retreat of moist vapor [the friend] would come up to my suite with a congenial young man and it got so that the house dick would follow him up to the suite to see where he was going, and make continual notes about it.

> I eventually noticed that I had begun to receive sour and disparaging looks from the management of the hotel, but this did not disturb me much for I had never gotten along well with managements or landladies, I mean not during my emancipated years.[50]

The Baron is the one character among the literary Romantics who does more than simply *talk* about love, about connecting and about making journeys. He does them all. While some will only see a self-loathing character (and behind him, a self-loathing author) who is no positive image for gay youth, it is also possible to see a young Tennessee Williams, stopping his bicycle along the Santa Monica palisade ("That depends on who's driving the bicycle, Dad"), searching for companionship and love, lighting a match to see into the eyes of a young man with whom he might share the night warmth and a warm bed.

Camino Real is far from a perfect play. Some would argue it is not even a particularly good one. Certainly, it is a difficult play to love, wearing, as it does, its oversized heart on its self-consciously colorful, calculatingly frayed sleeve. Its primary problem seems, at first glance, to be structural:

Camino Real is paralyzed between the story of the failed Romantics of the Siete Mares and that of Kilroy. The latter's story has to do with a kind of mendacity Williams saw creeping into American life, a cheapening of spiritual values, as personified in the Gypsy and her daughter Esmeralda, whose virginity is supposedly restored by each new moon, and whose encounter with Kilroy is anything but spiritual. Kilroy is looking for love and for a place where his dreams (whatever they may be) will be realized. What he finds is a street where his pocket is picked, his soul mocked, and his heart literally cut out. While Kilroy's story at least provides a colorful, active counterpoint to the static situation of the Romantics, and while it is thematically linked with theirs, there is no dramaturgical connection between them. One storyline has no effect on the other, and one could remove the Romantics from Kilroy's story and it would remain essentially unchanged, just as one could erase Kilroy from the Romantics' plot without visible dramatic result. Unable to choose between these two stories, Williams wrote a play without a center, in which Kilroy, who often seems to be the main figure, disappears for whole scenes at a time, then reappears in a series of episodes that, while freighted with symbolic meaning, meander and then dribble out.

Williams thought of *Camino Real* as a political statement, a comment on the social and political conditions of the United States in 1953. " . . . [E]ach time I return here I sense a further reduction in human liberties, which I guess is reflected in the revisions of the play," he told Henry Hewes, in an interview in *The Saturday Review* that March. In his article written for the Sunday *New York Times* the week the play opened, he wrote, "Of course, [*Camino Real*] is nothing more nor less than my conception of the time and world that I live in, and its people are mostly archetypes of certain basic attitudes and qualities with those mutations that would occur if they had continued along the road to this hypothetical terminal point in it."[51]

None of the daily critics mentioned the play's politics; mostly, they professed bafflement. But Barbara Baxley, who played Esmeralda, told Donald Spoto, one of Williams's biographers, that Walter Winchell and Ed Sullivan attacked the play as anti-American. "They accused it of being a leftist manifesto," she said. The play as performed in its pre-Broadway tryout in Philadelphia was much more explicit than it would be in New

York in its attacks on fascism, both in America and abroad. " . . . [A]ll references in the play to Fascism in America, and to brotherhood and love were cut, since they were thought to be ringing cries of Communist sympathy." Baxley doesn't say who thought so. It might have been the producers (who included Cheryl Crawford), or the director Elia Kazan, or Williams himself. Neither Kazan (who had testified as a friendly witness before the House Committee on Un-American Activities a year earlier) nor Williams mention this business of cuts in their autobiographies. In any case, not every such reference was cut: certainly, the forbidden word, *Hermano,* which causes Gutman such distress, is clear in at least one of its meanings.[52]

In *Camino Real,* Williams is indeed ready and willing to engage the political world, but only so far. The play catches Williams between his lifelong urge to flee, and the knowledge that it is sometimes better to stay and fight. He wrote, after all, in "The Catastrophe of Success" that man is born to struggle, and said in a 1972 interview, " . . . I've never stopped having to fight for existence[.] I think . . . that each play expresses a struggle to survive and be liberated." The play represents that no-man's land in Williams's psyche between Val Xavier's "Nobody ever gets to know *no body!*" and Lady's "NO, NO, DON'T GO . . . I NEED YOU!!!" in *Orpheus Descending.* To stay and fight, as Val does, is likely to mean pain and death. On more than one occasion, Williams said he was a revolutionary. But he was a revolutionary of a limited sort.[53] Speaking with an interviewer in 1975 about what he hated most about the early 1950s and McCarthyism, he said,

> "It's the moral decay of America, which really began with the Korean War. . . . We're the death merchants of the world, this once great and beautiful democracy. People think I'm a communist [for saying this], but I hate all bureaucracy, all isms. I'm a revolutionary only in the sense that I want to see us escape from this sort of trap."[54]

Williams's tendency toward flight usually won the struggle against any instinct to stay and right a wrong; his Dakin side usually prevailed over the fighting spirit of the pioneer Williamses. In the 1950s, as Williams looked around at America and saw the persecution of writers, blacks, Communists, and homosexuals, the part of him that would later declare himself to be the "founding father of the uncloseted gay world" may indeed have wanted to revolt. But this part of his nature warred with the part for which flight was instinctual. The instinct to revolt fought Williams's urge to flee to a stalemate in *Camino Real*, just as the stories of Kilroy and the Romantics battle each other to dramatic stasis. It is perhaps serendipitous that the stalemate reflected in the play happened to capture the American political moment as well as Williams's own condition.[55]

Camino Real is political in the same sense that Clifford Odets's plays of the Depression were political: it yearns for a better way, a more humane approach to the problem of living, but neither Odets nor Williams laid out any programmatic political beliefs. Williams was known to say on occasion that he was a socialist, as he does in his *Memoirs* and in some letters to Donald Windham. But, as Harold Clurman, who directed the first productions of many of Odets's plays wrote about that playwright,

> A tendril of revolt runs through all of Odets' work, but it is not the same thing as a consistent revolutionary conviction. . . . The "enlightenment" of the thirties, its effort to come to a clearer understanding of and control over the anarchy of our society, brought Odets a new mental perspective, but it is his emotional experience, not his thought, that gives his plays their special expressiveness and significance.[56]

The same can be said of Williams, and of *Camino Real.* The nature of Williams's "revolt," his occasional taste for revolution, is not problematic, but it is complex. Viewed through a political lens, for example, his decision to hire Kazan to direct *Camino Real* could be viewed as, at best, retrograde, a betrayal of those whom Kazan had betrayed. On the other hand, one can argue, as Kazan did, that the choice of Kazan as director so soon after his testimony was an act of brotherhood and loyalty to another who was an outsider. For Williams, however, politics didn't enter into the decision in the least. "I take no attitude toward [Kazan's testimony] one way or another, as I am not a political person and human venality is something I always expect and forgive," he wrote his friend Maria St. Just as he contemplated directors for *Camino Real* (he also considered the young Peter Brook and José Quintero). A life-long empathy with the oppressed and the suffering implied a political stance in Williams's work, but beginning as early as 1937 with *Spring Storm,* politics would always take second place to artistic concerns.[57]

━━━━━━━━

One way of viewing this dichotomy between the Romantics and the Gutmans/Prudences is to say that the Romantics are being punished for daring to love in a world of calculation, power, and double-cross. This is true, to the extent that their idealistic notions of love no longer have a place in a world in which everything is seen through the calculus of commercial value and political power. A less charitable way of viewing them is to suggest that they cannot face the new world, and have no weapon to oppose the powerful, offering only paralysis or flight, which are no weapons at all. Mere love of other human beings, Williams seems to say, is not armament enough to fight for the world in 1953. But this is as far as the play goes. Williams never directly says what is needed to oppose the Gutmans, the Trumans, and the Eisenhowers.

While the play lays out a bleak vision of the United States of 1953, it offers no way off the road that leads to a world of haves and have-nots, a world where the one fountain that still flows is guarded night and day to keep away the poor. It is a world populated by those who exercise power only to maintain their hegemony, and by a group of Romantics who yearn for something better but lack the will to fight, leaving the dispossessed and powerless to fend for themselves. The Romantics are far from being the world's salvation. The Baron, at least, is true to himself, and engages the rest of the world. In this sense, Williams's homosexual character is the one who actively believes, in 1953, in the democracy of sex, who acts boldly and publicly on that belief, and who, on account of it, is murdered. If one compares the Baron, who fits the more recent gay prescription of "out and proud" rather well, to the other gay or proto-gay characters of plays presented on Broadway in the same years—the evil Mr. Dulcimer (so evil that his motivating homosexuality must be denied in production); the guilty suicide Martha Dobie of *The Children's Hour;* the sensitive-but-not-gay Tom Lee and the self-loathing gay Bill Reynolds of *Tea and Sympathy;* and the swish stereotypes of *Season in the Sun*—it is difficult to sustain a charge of homophobia, heterosexist discourse, or closetedness against Williams. In the coming years, however, critics would continue to try their best.

In *Camino Real,* Williams doesn't offer a solution to the world's political and moral failures; he offers a photograph of it. In this, he follows Chekhov, who wrote Alexei Suvorin, "You are right to demand that an author take conscious stock of what he is doing, but you are confusing two concepts: *answering the questions* and *formulating them correctly.* Only the latter is required of an author." However, in *Camino Real* an ominous trait is emerging: one of an inner paralysis that will be reflected in some of Williams's work to come. This paralysis would carry over into *Cat on a Hot Tin Roof.*[58]

THREE

Something Kept on Ice

I

As Williams recovered from the poor reception of *Camino Real*, the campaign against gay men and lesbians in America broadened and intensified. Days after the opening, Williams and his lover, Frank Merlo, returned to Key West. Two months earlier, at the order of the mayor, Miami police raided several gay bars and cruising areas following the murders of two gay men who had been picked up in bars. What followed was a classic example of further victimizing the victims. The City Council passed a law requiring professional chaperones in movie theatres to protect teenage customers from homosexual predators; it approved legislation forbidding the serving or selling of alcohol to homosexuals. The mayor then called for a change in the law so that gay men and lesbians could be prosecuted under an existing white slavery act. By the end of 1953, the names of those arrested for "homosexual offenses" in Miami were made available to police departments across the southern part of the state.[1]

Settled in his studio in Key West, Williams understood what was happening. He had seen gay-bashing in Key West and Miami before. Three years earlier, he described the persecution of homosexuals in Key West in two letters to Paul Bigelow. On April 5, 1950, he described a number of incidents of harassment:

> Every girl in town has been booked on vagrancy. As far as I know, at this point we are the sole exceptions. Erna and her little family have fled from the Keys. They were refused drinks and told by the management (at an after-hours joint) that they were on the "undesirable list". The pianist at the Trade Winds was hauled in and Lyle Weaver, organist at the Bamboo was told to be out of town by the fifteenth. Michelle from Hector's is out on $250 bond. He is about the only one of the bunch that refused to plead guilty. Of course Pat, the Female Imp at the Cabana, was almost the first to get the ax. [. . .] It is funny, but also pretty frightening and disgusting as an example of how Fascistic little southern communities can become when they have a mind to. There are sixteen different kinds of "vagrancy" in Florida. All you have to do is walk down Duval after dark if they are out for you.[2]

A week later, he brought Bigelow up to date:

> I am inclined to go along with you in your analysis of the Key West Situation. I am told that these clean-ups nearly always coincide with the "fin de saison" and in this instance, along with a hotly contested local election. It doesn't make it much more pleasant, but for the moment, at least, the pressure seems to have eased up and Frank and I have not yet felt the heavy hand of the law. Michelle had something of a triumph in court, at least according to his modest standards. The policeman testified that he had picked him up for having "a sissy walk". The Judge said, "I think this boy is just too good-looking" and the case was dismissed without a fine being imposed. [. . .] The only unpleasantness we have had was in the Patio bar on the beach when a sailor came up to me and said, "Didn't Oscar Wilde smoke a cigarette holder like that?" I made big eyes and said, "Who is Oscar Wilde?" Social poise *always* comes in handy.[3]

As usual, Williams's response to discrimination would be restless artistry. He set to work revising *Camino Real* for publication before traveling to Houston to direct his friend Donald Windham's play, *The Starless Air.* Williams left Houston before the opening, drove to Memphis to see his grandfather, then by train went on to New York, whence he and Merlo sailed for Europe early in June. Following a brief stay in Rome, they traveled to Spain, then back to Rome, then on to Vienna, Venice, Zurich, Spain again, and southern Italy. By early autumn, Williams was in Tangier with Paul Bowles. At the end of October, he was back in New York.[4]

His habitual restlessness was now compounded by a serious inability to write. No place he visited proved any more conducive to work than the last, and as the weeks wore on, he became increasingly bad company for Merlo and anyone else who crossed his path. In mid-October, he wrote Audrey Wood that he had to drink in the morning before being able to write, but that the results were so poor that he would end the day more depressed than when he began. He was convinced that his problem was physical.

Back in New York, he moved into an apartment on East 58th Street near First Avenue. From there, Williams could have observed what was becoming a ritual in American politics: the election-year series of roundups, arrests, and gay-bashings. As a run-up to the mayoral elections of 1953, police raided gay bars and swept the dunes and beaches where gay men still gathered before Labor Day. Gay men were arrested by the dozens on Sutton Place, a popular cruising area minutes away from Williams's apartment. As usual, the names of those arrested appeared in the next day's newspapers.[5]

After a brief stay in New York, Williams pulled up stakes again, and with Merlo and the Reverend Dakin, who had joined them on their return from Europe, headed for New Orleans. Here, too, the gay population was under siege. The police recently had arrested 64 women at a lesbian bar in the French Quarter; the next day, the courtrooms were full of men and women arrested at other bars all over the city.[6]

In New Orleans, Williams finally discovered a physical ailment worthy of worry and surgery: hemorrhoids. He scheduled an operation at the Ochsner Clinic, the hospital that would play a pivotal role in his next play, but at the last moment, he had the minor procedure done at the Touro Infirmary instead: Ochsner had a firm policy against its patients drinking alcohol; Touro was more permissive on this point. Afterward, with Merlo and Grandfather Dakin still in tow, Williams headed back to Key West. In February 1954, Williams wrote to Donald Windham that his psychic condition was no better than it had been in Europe, and that he was doing little but going to the beach and the movies.[7]

That summer, he asked Cheryl Crawford to recommend a psychiatrist he could consult in order to get to the bottom of his writer's block. He changed his mind, however, and diverted himself with more travel and with the film version of *The Rose Tattoo* starring Anna Magnani, filmed in Key West (in which he and Merlo appeared briefly). And, slow and painful though it was, and with the consequences of being exposed as a gay man or woman displayed all around him, he started working on a new play, based in part on his short story of 1951–2, "Three Players of a Summer Game."

II

Meanwhile, Broadway was experimenting with tolerance. With the exception of *Camino Real* and Sartre's *No Exit,* produced in 1946, the plays that had dealt with homosexuality approached the subject either homophobically (*Season in the Sun*), as a false accusation (*Tea and Sympathy*), or both (*The Children's Hour*). Ruth and Augustus Goetz, the husband and wife playwriting team who in 1947 had a success with their adaptation of Henry James' *Washington Square* called *The Heiress,* decided in 1950 to adapt Andre Gide's novel, *The Immoralist.* From the beginning, the project was intended by the Goetzes as a protest against the treatment of homosexuals. Their outrage had been spurred by their friendship with John Gielgud, who had stepped in to replace the original director in the London production of *The Heiress.* The Goetzs' admiration for the sensitive, refined Gielgud, who had rescued their chaotic production and turned it into a hit, was boundless. They were aware that he was gay, and that he had to keep that part of his life out of public view. "We decided," Ruth Goetz said later, "for the sake of this fine, superior human being and others like him, we were going to drag the subject out into the open, out on the stage to which we had access since we were writing for the theater." (The Goetzs' concern for Gielgud was not misplaced. Three years later, he was arrested for "public indecency" with another man in a public lavatory in London).[8]

It took the couple ten months to find someone willing to produce *The Immoralist*; eventually Billy Rose took the chance. The production opened in February 1954, around the time Williams wrote Windham about trying to put himself back together in Key West after the failure of *Camino Real* and his prolonged inability to write. Freely adapted from Gide, *The Immoralist* is a curious and frustrating play. Michel, a young French archaeologist, impulsively marries to avoid admitting to himself his homosexuality. He takes his young wife on an expedition in North Africa, where, unable to consummate the marriage, he becomes increasingly tormented. There he meets a gentle Arab scholar, Moktir, who, rather than disguising his own homosexuality, has given up a prestigious professorship at the University of Fez. Michel is threatened by the Arab's openness, and more so by Moktir's ability to see that Michel is a homosexual, too. Moktir urges the Frenchman to be honest with his wife (who by now senses the unspoken truth and has taken to drink). Reluctantly, Michel confesses at the end of Act Two. "Listen to me, Marcie!" he cries. "You must go home hating me! For your own sake you must see me as I am." As if speaking of his homosexuality aloud at last has made him free, Michel goes on to declare a new openness: "I will never be silent again!" he promises. "Whoever knows me will know that about me first. Whoever hears of me will hear that before anything! If there is an ounce of energy within me, I will say what I am like! This one thing I can do! I can speak out!"[9]

Alas, Michel's determination doesn't survive intermission. The play's final scene takes place back in France, a mere six weeks later. Marcie has returned, pregnant with Michel's child, the result of the one night they managed to have sex. In the evening, as dark approaches, Michel appears at the house and asks his wife to take him back. The life of freedom and joy he had proclaimed a few weeks earlier didn't bloom as he'd envisioned. To the old family retainer, Michel describes the life he's led since Marcie came home: "I have scavenged through the back streets of every town from Tunis to Paris!" he says. "I've been exploited by those who are like me, and shunned by those who are not. . . . There is no loneliness like that!" (55). He and Marcie reach an understanding: He will come home to stay; he will avoid the life he recently embraced; and they won't speak of it again.

The Immoralist was offered by the Goetzes as a plea for tolerance, although in the last scene they backed away from the condition they were asking tolerance for. Michel decides the only way he can live happily is to suppress the very part of his existence for which just weeks earlier he was willing to sacrifice everything. Only Eric Bentley—himself gay and married at the time—called the writers to account in his review in *The New Republic:*

> Perhaps the honest ending would have been to let the husband stay with the wife, both of them knowing that there would also be young men. Is this more than the public of 1954 would take? Possibly; but a humanitarian playwright would be interested in putting the matter to the test. To write a didactic play is to suppose yourself ahead of the public and to suppose the public in need of your advice. A didactic playwright can write *only* plays that are more than the public will take.[10]

The Immoralist, mild, undramatic, and conventional as it was, proved, indeed, to be more than the public could take: It lasted 103 performances. While in many respects it faced the "problem" of homosexuality more squarely for most of its length than did *Tea and Sympathy* or the revival of *The Children's Hour,* the Goetzes, for all their good intentions, fell short of what Williams accomplished during the Baron's 15 minutes of life in *Camino Real* a year earlier. *The Immoralist* treats an audience to its pity, asks for their understanding for the plight of suffering homosexuals. Williams, in the person of the Baron, asked for neither pity nor understanding, but only for an acknowledgment that people such as the Baron exist. Williams simply shows us Charlus's life—indeed, the part of it most likely to offend—without explanation or special pleading. *The Immoralist* asks its audience to understand and tolerate behavior different from its own; Williams merely shows. Pleading for tolerance doesn't seem to occur to him. As Bentley pointed out in his review, even the tolerance for which the Goetzes, in all sincerity and with good intentions, pleaded, wasn't real. "Our public's motto is: tolerance—provided there is nothing to tolerate." While the Goetzes are quite clear in pointing out that Michel's torment is imposed by society, he gives in to it nonetheless, and vows to live a conventional life with wife and son. Beyond the role the Baron plays as a sado-masochist seeking a trick, Charlus is not tormented in the least. Michel craves homosexual companionship, but at the same time is terrified by it. The Baron accepts who he is, openly seeks gay companionship, lusts for it. When Moktir first challenges Michel to tell Marcie the truth, Michel declares he cannot: "It would be as if I were glad of the truth and I am not." (140) The Baron, on the other hand, *is* glad. He lives (and dies) for those moments when the nightingales sing in the mattress.[11]

The Goetzes, however, should not be faulted too much, even though they succeeded at less than they hoped or thought. They had just spent ten months searching for a producer willing to brave reviews such as John Chapman's: "'The Immoralist' . . . is a clinical drama about a sex deviate who manages one husbandly gesture. . . . I was embarrassed for [the actors] last evening. And for me, too."[12]

III

Brick seemed to be throwing his life away as if it were something disgusting that he had suddenly found in his hands. This self-disgust came upon him with the abruptness and violence of a crash on a highway. But what had Brick crashed into? Nothing that anybody was able to surmise, for he seemed to have everything that young men like Brick might hope or desire to have. What else is there? There must have been something else that he wanted and lacked, or what reason was there for dropping his life and taking hold of a glass which he never let go of for more than one waking hour?[13]

This description of the wealthy young planter, Brick Pollitt, from the 1952 story "Three Players of a Summer Game," could apply, without much exaggeration, to Tennessee Williams in 1953, as he wrestled in the Key West studio with the play he was trying to draw from it. The story is masterful in its simple, straightforward recitation of Brick Pollitt's mysterious collapse. Brick was an athlete at Sewanee who married Margaret, a New Orleans debutante, Mardi Gras queen, and scioness of the owner of a fleet of banana boats. Margaret seizes control of the Delta plantation, acquires Brick's power of attorney, and assumes the management of his business affairs, all the while insisting that her husband is merely passing through a phase. But Brick's business interests are not all Margaret seizes. In a passage of remarkable clarity and hardness, Williams describes the transformation that affects husband and wife:

> It was as though she had her lips fastened to some invisible wound in his body through which drained out of him and flowed into her the assurance and vitality that he had owned before marriage [. . . .] she became vivid as Brick disappeared behind the veil of his liquor. [. . .] She abruptly stopped being quaint and dainty. She was now apt to have dirty fingernails which she covered with scarlet enamel. (306)

She cuts her hair short. She has, in effect, replaced her husband in the male world of power with her own new self by inhaling his masculinity, leaving him as little more than an empty highball glass awaiting the next gin and tonic.

Brick's response is to increase his drinking and find a mistress. The woman is Isabel Grey, the wife of Brick's handsome young doctor. The doctor is afflicted with a terrible disease: "An awful flower grew in his brain like a fierce geranium that shattered its pot." As the tumor grows in size, it affects Dr. Grey's memory, speech and balance, until finally he is put to bed, where he lies, his eyes "blazing with terror." For a week, the young doctor is fed morphine through a hypodermic. During that week, Brick visits Isabel. With motives that are ambiguous and undescribed, Brick fills a hypodermic needle (his trembling hands suddenly become "sure and steady") and "pumped its contents fiercely into her husband's hard young arm." Dr. Grey dies; Brick slips between his sheets (307).

Just as Margaret has taken control of Brick's affairs, Brick assumes command of the widow's. He renovates and repaints her rundown house while the widow Grey and her daughter put up at a hotel at Brick's expense. The narrator, who at the time of the events was a child about the same age as Mrs. Grey's, tells us that while townsfolk initially applaud Brick's generosity, their approval quickly curdles into resentment. It becomes clear to all that Brick's reasons for assisting the widow are not philanthropic. Public opinion hardens into sympathy for Margaret—she who has put up with so much—and against Brick and Mrs. Grey, who had not lived in town long enough to make a good impression.

Brick is not deaf to the whisperings of his neighbors, and takes to defending himself at the top of his voice to the renovators working on the Grey house. He congratulates himself for "licking" his drinking problem, and berates Margaret for stealing his manhood. "Whatever I do," he declares, "I do it without any shame, and I've got a right to do it. I've been through a hell of a lot that nobody knows" (312).

But Margaret will be the victor in this game. Late one evening, as Brick is dashing across the wide, lush front lawn of the Grey house carrying an imaginary football, he trips over a croquet wicket and sprawls. He decides to cool himself off in the sprinkler on the yard, and, glass of gin in hand, strips down to his underwear and tie and wriggles beneath the refreshing water. Passersby gawk; eventually the chief of police arrives and gently takes Brick away. After that, Brick's visits to Mrs. Grey become sporadic. The widow soon sells the house and the rest of her possessions, and she and her daughter depart, leaving behind only their electric car and gossiping neighbors.

The narrator recalls seeing Brick one more time, on a brilliant autumn afternoon. Margaret is driving through town in Brick's Pierce Arrow touring car, while in the back, "pitching this way and that way with the car's jolting motion, like a loosely wrapped package being delivered somewhere," is Brick. Margaret waves at passersby and blows the car's silver trumpet at every intersection, calling out the names of people she recognizes. In the back seat, Brick sways, grinning with a "senseless amiability [. . . .] It was exactly the way some ancient conqueror, such as Caesar or Alexander the Great or Hannibal, might have led in chains through a capital city the prince of a state newly conquered" (324–5).

The story is one of Williams's best. His rigorous control of its polished, assured surface only hints at a rich subtext roiling beneath. Why did Brick begin drinking? What had caused his secret suffering? Why did he pump the hypodermic needle so powerfully into the dying doctor's arm? It was for lust and from need, but not merely for those things. Brick and Isabel know this, but neither one can say exactly what else was involved. They did not make love after watching Dr. Grey die: " . . . the only movement between them was the intermittent, spasmodic digging of their fingernails into each other's clenched palm while their bodies lay stiffly separate [. . .]" (307–8). As much as Williams reveals he conceals, and the result is Pinteresque in its silent mystery.

IV

As Williams fought his writer's block with work and alcohol, he transformed this story about marital infidelity and the mysterious transference of gender characteristics into a play about a man unable to acknowledge his homosexuality for fear of what the world might do with the news. In fleshing out the material from short story to full-length play and probing Brick's psyche, which earlier he had left untouched, Williams added key elements that complicated the play considerably—seemingly as much for himself as later for director, audience, and critics.

Perhaps in raising the issue of homosexuality and how a gay man could be paralyzed with the fear of the very real consequences of exposure, he was reacting to the events occurring around him in Miami, New York, New Orleans, and elsewhere. At the same time, he was reaching back into his days as an unattached and recently out gay man. In Provincetown in 1941, a young woman convinced Kip Kiernan that Williams was trying to convert him into a homosexual, and Kip withdrew. Williams threw a boot at the head of the interfering woman. In *Cat on a Hot Tin Roof*, Williams rearranges the scene. Maggie recalls to Brick how she told Skipper, "STOP LOVIN' MY HUSBAND OR TELL HIM HE'S GOT TO LET YOU ADMIT IT TO HIM!" and Brick hurls his crutch at her head. In trying to make Skipper (and perhaps Brick) face the truth, Maggie creates a situation far worse than did her real-life counterpart: Kip broke off with Williams, and the playwright, although traumatized and broken-hearted, moved on. As a result of Maggie's insistence on truth, Skipper slides into alcohol and suicide and Brick is seized by a paralyzing guilt.[14]

When Williams fled to Mexico in the wake of Kip's crushing rejection, he met a young man who wanted to have an affair with him. He described the man to Donald Windham as a wealthy former football star for the University of Michigan who now did little but talk of suicide. This ex-football star may well have been a model for Brick Pollitt in the short story, but there Williams submerged the man's homosexual and suicidal feelings. As he worked on the play, he resurrected both, and now traces of the football star are found not only in Brick, but in his athletic companion Skipper, whose name bears more than a faint echo of the dancer's for whom a younger Williams fell so hard, and by whom he'd been so badly hurt.[15]

Williams's heart was not nearly as calcified as he both hoped and feared, and Kip was not so easily forgotten (for years Williams carried a photo of him and reproduced it in *Memoirs*). Nor did the hurt Kip inflicted fully heal. Both would be worked over in the new play. Perhaps through Dr. Grey and Skipper, Williams was venting anger, exercising a psychological revenge on Kip Kiernan. Kip died of a brain tumor; Dr. Grey dies of a brain tumor. Skipper will die a death that Williams doesn't specify, but is a projection of Williams's own worst fear: dying alone and rejected.

If Williams was exacting a psychological revenge on Kip in "Two Players of a Summer Game" and *Cat* by killing him off, it was probably an unconscious one, and all the more damaging, psychologically and dramaturgically, for being psychically unavailable. It also probably filled him with guilt. For as deeply as Williams was wounded by Kip, he continually punished himself for transforming him, and others (his sister Rose especially), into material he could exploit in his work. Williams hated using other people's misfortunes as material, but as a writer had no other choice. As late as 1981 in *Something Cloudy, Something Clear* and *The Notebook of Trigorin,* he would confess his guilty feelings for using people, and try to come to terms with them. Being used by her husband as mere material is Zelda Fitzgerald's accusing refrain toward Scott in *Clothes for a Summer Hotel.* Catharine Holly will allude to such guilt in a famous line in *Suddenly Last Summer,* which, written two years after *Cat,* is a meditation on the ways people use each other and call it love.

Brick's guilt and spiritual paralysis, then, seem to spring as much from Williams's own guilty feelings—not over homosexuality, but concerning his appropriation of other peoples' lives for his own art and enrichment—as from anything in Brick's fictional history. This complication, like Williams's conflicting emotions about Kip, originating as it does outside the play, causes *Cat's* dramaturgy to be murky at best. Hence the hole in the play's fabric that has so long frustrated and puzzled actors, directors, and critics.

Laying across this hole, as it were, is another layer of guilt, and this one *will* have to do with homosexuality. But that guilt has nothing to do with *being* homosexual; rather, it stems from the reluctance to admit it to the world, because the world in 1954 was so quick to punish the admission. For a writer whose self-appointed task was always to tell the truth, such a lie, clashing with the desire *not* to lie, could be paralyzing, indeed. Thus, in approaching this dangerous material, Williams was caught in a double bind, trying to reconcile in the play not one but two psychic conflicts.

Williams's own life was becoming severely disrupted by the alcoholic symptoms he described in Brick. As those symptoms began exhibiting themselves in 1953, Williams returned to their fictional manifestation in a story from a year earlier and begins to puzzle out their source. The original short story works as well as it does because its subtext is so thoroughly submerged—as unknown to Williams, perhaps, as to the reader. In the play, where Williams wants to be at least as concerned with the causes of Brick's drinking as with its effects, the unconscious material begins to surface and this is where his troubles begin: Williams creates an offstage character whose name is unmistakably close to Kip Kiernan's, and includes in Brick's portrait details from the sad life of the former football star with whom Williams was briefly involved while he was attempting to get over Kip. The material, shot through with his guilt for using it, is still too fraught for Williams to control artistically. Unlike Brick, who needs to have a drink in his hand at all times, Williams, at least according to the letter he wrote to Audrey Wood, needed to drink only in order to write. Whatever was troubling him enough to make him turn seriously to alcohol was a matter that expressed itself, albeit incompletely and confusedly, in his work.

The composition of the play was made yet more difficult by the famous concerns regarding its dramaturgy of the director, Elia Kazan. Kazan, on reading the draft that Williams sent him in the spring of 1954, was interested, but had reservations. As Williams summarized them in a "Note of Explanation" first published in the 1955 mass-market paperback edition of the text, Kazan wanted a Maggie who was softer and whom an audience would find more sympathetic than she was in the original draft (in which she seems closer to the ruthless and masculine Margaret of "Three Players of a Summer Game"). Kazan thought Big Daddy "too vivid and important" to disappear from the play after Act II, and he wanted to see some change in Brick following his "virtual vivisection" by Big Daddy. Williams wrote in his Note that while he complied with all three of Kazan's wishes, he was comfortable only with the first. Nonetheless, he made the requested changes because he had confidence in Kazan and, despite the failure of *Camino Real,* wanted him to direct. Williams felt at least partially justified in his acquiescence, for the rewrites helped the play to win the Pulitzer Prize. "The reception of the playing-script," Williams wrote in the Note, "has more than justified, in my opinion, the adjustments made to [Kazan's] influence. A failure reaches fewer people, and touches fewer, than does a play that succeeds" (125). Williams, however, was also of another, less sanguine mind regarding Kazan's influence. In the mass-market edition, published by Signet shortly after the play opened, Williams included the original Act III, in which Big Daddy does not reappear and Brick does not undergo a significant change resulting from his Act II confrontation with Big Daddy. This was followed by the "Broadway Version," written at Kazan's behest. That Williams's original version comes first, followed by the explanatory Note and then the revised Act III, strongly suggests the belief he still had in the first version, as does the decision to publish both in the first place.

What Williams leaves out of his "Note of Explanation" is any mention of the numerous changes he made, passages he cut from the production and then restored in subsequent published editions, concerning the relationship between Brick and Skipper. Williams was a compulsive rewriter; but even for him, the number of cuts, reinstatements, and re-excisions in *Cat* is high. They suggest that in the matter of Brick and Skipper, Williams was unusually uncertain and indecisive, and that he remained so long after it came time to publish the play. In the Dramatists Play Service acting edition of the play, published in 1958, he made yet more revisions, but did not print the original version of Act III or his "Note of Explanation." In the version he published in 1971 as part of his collected plays, *The Theatre of Tennessee Williams,* still more revisions

appear, along with the Note and the original third act. In 1975, he published yet another version based on a 1974 production at the American
Shakespeare Theatre. This one, published without the "Note of Explanation," contains a third act that combines elements of both.

The play has two primary lines of action. One has to do with
telling and facing the truth; the second, with maintaining one's place
in the world. Big Daddy must come to terms with his mortal illness;
Brick must admit to himself his culpability in Skipper's death and,
perhaps, his homosexuality. In doing so, Brick also confronts his desire
to retain what Williams calls in a significant stage direction, his "early
laurel" (89). The rest of the family is also caught up in a fierce struggle
to inherit Big Daddy's vast holdings upon his increasingly imminent
demise.

In the pre-production script that Kazan approved in November 1954,
Williams gave a speech to Maggie that described Brick and Skipper's relationship during their college days at Ole Miss, which, while not mentioning a sexual relationship, does indicate a closeness that makes Brick
vulnerable to a charge of a love by no means platonic. Williams cut the
speech during rehearsals, and it did not appear in the production. He restored some of it, however, in the Signet edition and retained it in the
1975 New Directions American Shakespeare Theatre edition:[16]

> It was one of those beautiful, ideal things they tell you about in the Greek
> legends, it couldn't be anything else, you being you, and that's what made
> it so sad, that's what made it so awful, because it was love that could never
> be carried through to anything satisfying or even talked about plainly.
> Brick, I tell you, you've got to believe me, Brick, I *do* understand all about
> it! I—I think it was—*noble!* My only point, the only point that I'm mak-
> ing, is life has got to be allowed to continue even after the *dream* of life
> is—all—over. . . . Why, I remember when we double-dated at college,
> Gladys Fitzgerald and I and you and Skipper, it was more like a date be-
> tween you and Skipper. Gladys and I were just sort of tagging along as if it
> was necessary to chaperone you!—to make a good public impression—
> (43–4)

At which point Brick threatens to hit Maggie with his crutch. When
Williams published the play for a second time, in the Dramatists Play
Service acting edition, in 1958, he cut the details about Brick's college

dates again, only to restore them in the 1971 collected plays edition published by New Directions.

Also included in the 1955 published text but omitted from the 1958 edition and restored in 1971 was most of a speech by Maggie in which she explained why she and Skipper had sex one night in a hotel in Chicago, while Brick was hospitalized with a football injury. It was earlier that day that Maggie had urged Skipper to "STOP LOVIN' MY HUSBAND OR TELL HIM HE'S GOT TO LET YOU ADMIT IT TO HIM!," after which, according to Maggie, Skipper "ran without stopping once, I am sure, all the way back into his room at the Blackstone . . ." (45). That night, Maggie came to his room and in an attempt to prove to her that her accusation was untrue, Skipper tried unsuccessfully to have sex with her: "And so we made love to each other to dream it was you, both of us!" (42–3).

Maggie recounts these events, which Brick can barely stand to hear, because she is trying to convince him of the importance of honesty. She tells him not the truth only about Skipper, but also regarding her desire to have sex with Brick again. She loves him, but after all, in order to have a chance to inherit the plantation, they ought to have children, just like Gooper and Mae, who now are expecting number six. She is equally frank about her desire to inherit the plantation. All of this, she tells Brick, she can at least admit honestly. One of the reasons she admires Big Daddy is, "Because Big Daddy *is* what he *is,* and he makes no bones about it" (41).

Here the play's two lines of action intersect, for if Brick is to admit, like Big Daddy, who and what he is, he will have to come to grips not only with his culpability in the matter of Skipper's death, but also with the nature of their relationship and its potential social consequences. Like Maggie, Brick seems generally in the habit of telling the truth. He won't lie to Big Daddy about forgetting his birthday; he doesn't for a moment try to hide his alcoholism from anyone in the family or their friends who have gathered for the party. But he cannot tell the truth when it comes to his relationship with Skipper, because doing so requires facing the world's disapproval.

Williams was similarly uncertain about how much he wanted to say in the long scene in Act II in which Big Daddy, shocked by his son's condition, tries to ascertain its origin. Big Daddy is a tolerant man: "Always, anyhow, lived with too much space around me to be infected by ideas of other people. One thing you can grow on a big place more important than cotton!—is *tolerance!*—I grown it," he tries to reassure Brick (89). Two people who did infect Big Daddy with some ideas were the owners of the plantation when he arrived there as a young man, broke and shoeless: Jack Straw and Peter Ochello, who shared for years the very room and bed where Brick and Maggie are no longer lovers. In a draft that preceded the version with which Williams and Kazan began rehearsals, Big Daddy spoke with admiration of the two men's love and loyalty in a long speech, in which he compared their behavior to Brick's refusal to speak to Skipper when the latter called him drunkenly to finally confess his love. While Brick upholds his version of their relationship as pure because it was platonic, Big Daddy believed it was a lesser love than that of Straw and Ochello, which was based on loyalty and honesty. But Williams cut this speech before rehearsals began. He restored a small portion of it in the 1955 published text (and retained it in later editions), describing Big Daddy's arrival at the plantation, how the pair of owners took him in, and, "Why, when Jack Straw died— why, old Peter Ochello quit eatin' like a dog does when its master's dead, and died, too!" (86)[17]

In the pre-production script, Brick's response to Big Daddy's mere mention of Straw and Ochello in the same breath with his own and Skipper's is fury. He fairly shrieks at his father,

> You think so, too? You think so too? You think me an' Skipper did, did, did!—*sodomy!*—together?
> Big Daddy: Hold—!
> Brick: That what you—
> Big Daddy:—ON—a minute!
> Brick: You think we did dirty things between us, Skipper an'—[. . .] You think that Skipper and me were a pair of dirty old men? [. . .] Straw? Ochello? A couple of—
> Big Daddy: Now just—
> Brick:—ducking sissies? Queers? Is that what you—? (87–8)

Williams cut most of this for the production, but restored it in the published texts. In Williams's original Act III, there is no indication that Brick has undergone any change from this fierce colloquy. At Kazan's request, however, he suggests a provisional possibility of a change. Brick says to Maggie near the opening of the "Broadway Version" of Act Three, "I didn't lie to Big Daddy. I've lied to nobody, nobody but myself, just lied to myself. The time has come to put in me Rainbow Hill [a drying-out center for alcoholics], put me in Rainbow Hill, Maggie, I ought to go there" (127).[18]

What was it, however, that Brick has been lying about? His failure to allow Skipper to be honest with him about his love? His own erotic love for Skipper? Guilt about both? Williams is not explicit. He allows both possibilities to be subsumed under the general topic of loyalty. This sudden vagueness leaves us feeling dissatisfied and, perhaps, cheated.

V

While giving the play the best notices Williams had received since *Streetcar*, the critics tended to share none of Williams's uncertainty regarding the relationship between Brick and Skipper. Robert Coleman described it as "unnatural"; John Chapman called Brick "a drunkard and queer," and felt "frustrated . . . some heart or point or purpose was missing." Richard Watts wrote that Brick "has a deep terror that he is homosexual," and John McClain that he "finds himself unable to rid himself of an infatuation for his college room-mate . . . and is hence incapable of either [a] normal relationship with his wife or any protracted periods of sobriety." William Hawkins alone thought that only "in the end is the truth of the young men's puzzling relationship clarified."[19]

Walter Kerr paused to wonder more deeply about Brick and Skipper, and he echoed Chapman's suspicion that something was missing. "'Cat On a Hot Tin Roof' is a beautifully written, perfectly directed, stunningly acted play of evasion: evasion on the part of its principal character, evasion, perhaps, on the part of its playwright," read the first paragraph of his review. "Brilliant scenes, scenes of sudden and lashing dramatic power" could not divert Kerr from a nagging feeling that the playwright was keeping too much to himself. It is worth quoting at length:

There is, however, a tantalizing reluctance—beneath all the fire and all the apparent candor—to let the play blurt out its promised secret. This isn't due to the nerve-wracking, extraordinarily prolonged silence of its central figure, to his merely repeating the questions of other people rather than answering them. It is due to the fact that when we do come to a fiery scene of open confession—between a belligerent father and his defiant son—the truth still dodges around verbal corners, slips somewhere between the verandah shutters, refuses to meet us on firm, clear terms.

We do learn, in a faint echo of "The Children's Hour," that there has been something to the accusation—at least on the part of Brick's friend. We learn that Brick himself, in his horror at the discovery, has done the damage he blames on his wife. But we never quite penetrate Brick's own facade, know or share his precise feelings. . . . In "Cat On a Hot Tin Roof" you will believe every word that is unspoken; you may still long for some that seem not to be spoken.[20]

Williams was distressed by Kerr's notice. Kerr was not only an influential critical voice, but also Williams's most appreciative and perceptive critic among the daily reviewers. Williams wrote a long reply called "About Evasions." Rather than mailing it to Kerr, however, he sent it to his friend Maria St. Just for, in her words, "safekeeping." Why Williams did not send it to Kerr is a small mystery. He might have feared Kerr would take offense, or he simply might have been as uncertain of the critic's reaction as he was on the matter of Brick and Skipper. He also may have feared that Kerr's suggestion that a vital, clarifying truth was missing was correct.

"[. . .] I still feel that I deal unsparingly with what I feel is the truth of character," he wrote. "I would never evade it for the sake of evasion, because I was in any way reluctant to reveal what I knew of the truth." He then quotes, with some minor changes, the long stage direction that appears in the 1955 published text (deleted for the 1958 acting edition), just as Big Daddy and Brick approach the cause of the young man's drinking. This is its crucial passage:

> The thing they're discussing, timidly and painfully on the side of Big Daddy, fiercely, violently on Brick's side, is the inadmissible thing that Skipper and Brick would rather die than live with. The fact that if it existed it had to be disavowed to "keep face" in the world they lived in, a world of popular heroes, may be at the heart of the "mendacity" that Brick drinks to kill his disgust with. It may be at the root of his collapse. Or it may be only a single manifestation of it, not even the most important.[21]

He then addresses Kerr's question about Brick directly:

> Was Brick homosexual? He probably—no, I would even say quite cer-
> tainly—went no further in physical expression than clasping Skipper's
> hand across the space between their twin-beds in hotel rooms and yet—
> his sexual nature was not innately "normal." Did Brick love Maggie? He
> says with unmistakable conviction: "One man has one great good true
> thing in his life, one great good thing which is true. I had friendship with
> Skipper, not love with you, Maggie, but friendship with Skipper. . . ."
> [. . .] But Brick's overt sexual adjustment was, and must always remain, a
> heterosexual one.[22]

Probably straight, no, certainly straight, yet "not innately 'normal'." Brick's "overt" sexual adjustment "must always remain" heterosexual. This declaration is less than straightforward; and "overt
adjustment" is not the same as "sexual orientation." Perhaps the overt
adjustment must always remain heterosexual in order to disguise another, shadowy truth: that Brick is an invert, perhaps an unknowing
one. In case Williams has not complicated matters enough, next he invokes Pirandello, who "devoted nearly his whole career as a playwright
to establishing the point that I am making in this argument: That
'Truth' has a Protean nature, that its face changes in the eyes of each
beholder. Another good writer once said, 'Truth lies at the bottom of a
bottomless well.'" Next, he adds an addendum: "The story must be
and remained [*sic*] the story of a strong, determined creature (Life!
Maggie!) taking hold of and gaining supremacy over and converting to
her own purposes a broken, irresolute man whose weakness was imposed on him by the lies of the world he grew up in." This last is odd.
While Margaret's appropriation of Brick's strength is a major event in
"Three Players of a Summer Game," Williams, seemingly under
Kazan's influence, muted the question of Maggie's "masculinity" in the
play, and her "taking hold and gaining supremacy over" Brick is over-
shadowed by Brick's paralysis and Big Daddy's attempt to get at its
cause. After spending so much time inhabiting both story and play,
Williams may honestly have thought that what was in the one was in
the other. And just what is the "weakness" imposed on Brick by the
lies of the world?[23]

Unlike critics, artists are rarely interested in putting their work in neatly labeled boxes. This may explain Williams's constant hedging over Brick's supposed heterosexuality. His hedging, however, has the feeling of genuine uncertainty. Nonetheless, in the midst of all of Williams's hesitancy, he says one thing quite clearly, first in the long stage direction, and then again toward the end of his letter to Kerr: It is not Brick's sexuality per se that may be "at the root of his collapse." Rather, it is, "The fact that if it existed it had to be disavowed to 'keep face' in the world they lived in [. . . .]" Brick's weakness was "imposed on him by the lies of the world he grew up in." While Brick's denial of his authentic sexual identity, too, proves to be negotiable (it *may* be at the root of his paralysis, or it may be only a single manifestation of it, "not even the most important"), Brick's wish to keep his position in the world is worth our attention if only because Williams brings the subject up twice.

As much as he abhors and denies the charge (made, Big Daddy tells us, by Gooper and Mae) that he is a "fairy" and breaks into a noticeable sweat over it, Brick is even more concerned about what people would think if it were true. "—Don't you know how people *feel* about things like that?" he demands of Big Daddy. He continues:

> How, how *disgusted* they are by things like that? Why, at Ole Miss when it was discovered a pledge to our fraternity, Skipper's and mine, did a, *attempted* to do a, unnatural thing with—We not only dropped him like a hot rock!—We told him to git off the campus, and he did, he got!—All the way to—[. . .] North Africa, last I heard! (88–9)

In Act I, Maggie addresses the same question while dodging Brick's threatening crutch. When Brick refers to the "one great good true thing" in his life, namely his "friendship" with Skipper, he accuses her naming that thing dirty:

> Margaret: I am not naming it dirty! I am naming it clean!
> Brick: Not love with you, Maggie, but friendship with Skipper was that one great true thing, and you are naming it dirty!
> Margaret: Then you haven't been listenin', not understood what I'm saying! I'm naming it so damn clean that it killed poor Skipper!—You two had something that had to be kept on ice, yes, incorruptible, yes!—and death was the only icebox where you could keep it. . . . (44)

Williams's part regarding his own father). So, too, is Maggie's determination to live—and her willingness to lie to keep on living. Those who, years after the fact, bemoan Williams's "failure" to make more clear the issue of Brick's sexuality and Brick's shame for not admitting it, misunderstand not only the complexity of Williams's feelings (simplifying them with the demands of a political slogan: "the positive image"), they also neglect the simple fact about the world in 1955: It was one thing for Williams to be out to his friends, family, and colleagues; it was another to make an announcement to the world. But while Williams made no announcement, neither did he ever take an action that could be considered a denial. He never married, and he never made a secret of his late-night carousings.

Certainly, that he was using Kip's rejection, suffering, and death as mere material was a truth he was uncomfortable with, as was the fact that he enjoyed his celebrity and the financial rewards his appropriation of the pain of others brought him. He may have feared, if he was entirely honest about homosexuality, losing the fame and wealth that came with being his country's leading playwright. Williams also felt himself in a perilous position as a commercial playwright (the only sort of playwright there could be in America in the 1950s): Three of his last four plays—*Summer and Smoke, The Rose Tattoo,* and *Camino Real*— had been financial failures. He had written candidly about what critics would consider the "sordid" aspects of gay life in short stories, but stories have audiences of one reader at a time, and in the 1940s and 1950s they received nothing like the attention given a Broadway play by an established writer; stories provided far less opportunity for a career-damaging scandal. For a playwright as well as for an audience, a play is likely to be a more threatening aesthetic experience than a story or a novel, because it is a live representation, which suggests much more strongly than the written word a concrete reality: The thing being enacted before us is far more likely to seem real than words on a page. It *is* real, if only as a representation. On the other hand, if Williams had had no qualms about concealing the truth as he knew it—about his homosexuality and the ways in which society condemned homosexuals—he could have written play upon play with ease; he might have been a boulevard playwright, genteel and discreet, who kept his secrets among an antique-filled apartment, who never mentioned the forbidden subject in his work, and blunted his torments with silence and alcohol. Williams endured his torments with increasing amounts of alcohol, and soon enough with drugs, but concealment was not one of his attributes—at least when not balanced by the equally strong urge to reveal.

VII

Life for gay men and lesbians in America in between 1954 and 1956 was scarcely different, and certainly no safer, than it had been before. True, as Williams worked fitfully on *Cat,* the power of Joe McCarthy came shuddering to an end, but McCarthy's eclipse, and the gradual waning of HUAC's influence, did not end the epidemic of homophobia. In November 1955, after a young boy in Sioux City, Iowa, was kidnapped and killed, the county attorney ordered the detention of all known homosexuals. Twenty-nine of them were committed to asylums without hearing or trial (prefiguring Williams's next play with a gay character, *Suddenly Last Summer*). Almost simultaneously, in Boise, Idaho, three men were arrested, charged with having sex with teenage boys. An investigation lasting over a year followed, and 1,400 residents were called into the police for questioning and were pressured to name homosexual acquaintances. Hundreds fled the city.[28]

Even Alfred Kinsey found himself in trouble. Since his first study appeared in 1948, he had been under constant attack from conservative members of the psychoanalytic community, particularly regarding his findings on homosexuality. A leading conservative analyst, Edmund Bergler, virtually accused him of being a traitor to his country. In a book called *The Homosexuals: As Seen by Themselves and 30 Authorities,* he wrote, " . . . Kinsey's erroneous conclusions pertaining to homosexuality will be politically and propagandistically used against the United States abroad, stigmatizing the nation as a whole in a whisper campaign. . . ." In 1954, a year after Kinsey published his volume on female sexuality, a special committee of the House of Representatives investigated charges that Kinsey's research undermined the American family and supported Communism. His principal patron, the Rockefeller Foundation, came under considerable political pressure to discontinue its funding of Kinsey. Later that year, it did.²⁹

It may have occurred to Williams that even a person as famous as himself, working in as relatively sheltered a place as the theatre, was not safe creating a gay character and placing him at the center of a play. This should not be confused, however, with the moral cowardice with which critics writing from the safe distance of the 1980s and '90s have charged him. He was engaged in an internal struggle in which his desire to write truly was pitted against real penalties for doing so. But he did not give in; he struggled against the feelings of mendacity that could overcome him when he was not writing freely. In this sense, Brick's struggle with his guilt for not admitting his love for Skipper may well have echoed Williams's own emotions about not writing what he knew to be the truth. In not confessing his true feelings for his teammate, Brick was also engaging in that other activity that caused Williams so much pain: By not recognizing Skipper as his lover, Brick in a sense was using him in the cause of his own advancement.

At the beginning of 1956, Kazan asked Williams to come to Mississippi, where he was preparing to shoot Williams's screenplay *Baby Doll.* Williams was reluctant to go because of the homophobia he often felt in the South. "Those people chased me out of there," Kazan recalled Williams telling him. "I left the South because of their attitude towards me. They don't approve of homosexuals, and I don't want to be insulted." Even someone as insulated from bigotry and hatred as Williams knew of their existence firsthand, and chose not to confront them alone. If this was a mistake, it was one he would have to grapple with in his work.³⁰

VIII

The problem Williams built into *Cat* with his failure to clarify the nature of Brick's sexuality and his feelings about it has provided critics with endless opportunities to engage in enthusiastic spadework, much of which has more to do with the political imperatives of their own time and place than with Williams or with Williams's concerns and experiences as a human being and an artist. John Clum, writing with the safe outrage possible in the early 1990s, affirms that homophobia clear and simple is

at the root of Brick's "sexual and emotional malaise," while the play's language "is no more positive than the heterosexist discourse that pervades
Williams's other work."[31]

As drama and as theatre, *Cat on a Hot Tin Roof* is the story of human
beings contending with competing emotions and desires, written by another human being engaged in his own struggle for something of unspeakable importance: a resolution in the battle between living an
authentic life and keeping the material comfort and fame that may be acquired if one fudges. In the world in which playwrights live and work, intricate, subtle, and often unconscious transactions between that life and
work are the immediate reasons that lay behind a piece's aesthetic failure
or success. In the case of Brick Pollitt, one might claim that Brick fails
ethically because, finally, he seems willing to deny his nature and a lover
before forfeiting his comfortable position in the world, and will not
admit that he is making this choice. One might posit next that the reason
the matter is so unclear and so unsatisfactorily dealt with is due largely to
Williams's own unresolved conflicts over this issue and over of his use of
other people in attaining success. In this conflict, politics and culture
clearly played an important part. But they are not theoretical politics;
they are personal and public politics as practiced every day, and which
Williams read about in the newspapers and magazines, saw in bars and
on the streets, and then mixed, with a degree of consciousness finally un-
knowable, with his own preoccupations, desires, and fears. It is, after all,
the unconscious element that separates art from propaganda, artistic
from purely political statements.

In the end, what do any of these considerations—biographical, historical, psychological, or the purely theoretical—have to do with the actor?
The actor playing Brick is forced to make choices that can rely only on the
actions and the circumstances provided by the playwright between the
covers of his play. Is Brick gay? Is his sexual nature more or less important
than his fear that he may be perceived as gay? An actor must choose, for
Brick's relations to Maggie, his father, and Skipper are dependent on
whom he has feelings for, the nature of those feelings, and with whom he
wants to share a bed. What can explain his paralysis? A man who is truly
comfortable with his sexual nature, and especially one as well situated in
life as Brick, has nothing to fear in the opinions of other people: He

knows who he is. A comfortably heterosexual man could tell his friend Skipper that he couldn't return his feelings and that, although painful, would be that. A gay man who loved Skipper as Skipper loved him has a choice. Either he can find the strength of character to admit it to himself and to Skipper, and then deal with the consequences; or, unable to admit that he loves a man, can cut the man off in the time of his greatest need. Then he can face the consequences of *that* act, or deny them. The few actions that Brick does take indicate that his regard for Skipper was not just that of the average heterosexual male. Nor does anything in the play suggest that Brick can admit to loving Skipper. What choice does this leave the actor? The only one that explains what Brick does (drink, evade, lash out) and what he doesn't do (sleep with his wife, admit that Skipper loved him, keep his composure in the face of his father's love and understanding) is that he did indeed love Skipper, but could not admit it to him or to himself. Brick, then, is gay, and for fear of losing his privileged place in life, cannot say it, perhaps even to himself. (Yet another choice the actor must make: If Brick has homosexual feelings, how aware of them is he? If he is unaware of them, can they be acted?)

The actor cannot afford to adopt Williams's own uncertainty on this matter. The next question for the actor, then, is, how to play Brick's dilemma? The answer may sound like an equivocation, but it is not. Rather than assume, as he usually would, that a character has one overall objective and that everything he does is an attempt to achieve it, the actor playing Brick may have to hold in his head two opposing objectives, one playing the major role at one moment, and the other the next. The one would be to tell the truth about his feelings for Skipper (as he does when it comes to other subjects); the other to protect the position he has attained in life. It is the clash of these two opposed desires that produces his paralysis. Another possible overall objective for Brick might be, "to find peace." Unfortunately, neither alternative grants him that— and again, paralysis is the result.

─────────────

Work only seemed to get harder for Williams. Later, he would say that the rewriting of the last act of *Cat* damaged him fundamentally as a writer. Perhaps it did; but he would also say that he never blamed Kazan for his troubles that followed that play, and by publishing both the original Act III and the one rewritten at Kazan's request, Williams was telling future producers that they could choose between them. In any case, after the premiere of *Cat,* he again experienced a prolonged period where he couldn't write, and from that point on, he would be largely unable to work except under the influence of drugs as well as alcohol. As usual, he traveled—to Key West, to Rome, to Barcelona, Tangier, New Orleans, New York, back to Key West. But he would be unable to escape a growing sense of constriction that would manifest itself in attacks of claustrophobia, obsessive rounds of travel, and increasing reliance on Seconal, which he referred to, benignly, as "pinkies." Whatever he was fleeing accompanied him from stop to stop, drink to drink, pill to pill. But then, on some level, he knew that. "One's enemy," he wrote to Maria St. Just in 1956, "is always part of oneself."[32]

FOUR

A True Story of Our Time

I

Williams had thought about entering analysis as early as 1954, when he was working on *Cat on a Hot Tin Roof.* He asked Cheryl Crawford for a referral, but if she gave him the names of any therapists, he never followed up. The fact that he was feeling sufficient psychic pressure to ask for help as he'd begun working on the play is significant in itself for what it says about his anxieties concerning the work's materials; that he put off such help is also significant. Perhaps he thought he could overcome any anxiety through writing; perhaps he was afraid of what he might discover about his subject matter should he probe it directly, without the intervening mask of imagination and symbol. By the beginning of January 1957, however, as preparations for the Broadway production of *Orpheus Descending* were getting underway, he wrote Sandy Campbell, Donald Windham's lover, that he planned to start psychoanalysis as soon as he returned to New York from Key West, at the end of the month. But rehearsals for *Orpheus* may have interfered. It was not until after the twin disasters of poor reviews of *Orpheus Descending* in the daily papers of March 22, and his father's death five days later, that Williams went for the help he believed psychiatry could offer.[1]

There were several reasons why the negative reviews of *Orpheus Descending* and its quick closing, after 68 performances, were so devastating. Williams was growing more sensitive to any sign of critical or popular disapproval of his work; a year later, still smarting from the play's quick demise, he told an interviewer that his best work was behind him and that the public was no longer interested in his work. A few days after that interview, however, he would unleash on that public a play even stranger and more violent than *Orpheus Descending*.[2]

More significant, perhaps, was his lack of success—for the second time—with the very personal material of *Orpheus*. The story of Val Xavier and Lady Torrance, of the young, vulnerable artist as outsider, was important enough to Williams that he had never abandoned it, even after its Boston debacle as *Battle of Angels*. The play, he said, had never been off his workbench. The closing of *Orpheus Descending* no doubt not only reminded him of that painful earlier experience, and of his abrupt dismissal, as it seemed to him at the time, by the play's producer, the middle-brow, stuffy Theatre Guild. In truth, *Orpheus Descending* is a significantly better play than *Battle of Angels*. Its conflicts and actions are clearer, more compelling, and move with far more assuredness and inevitability, but only when performed according to Williams's explicitly expressionistic stage directions. The 1957 production, directed by Harold Clurman, was decidedly realistic, and what the audience saw was lurid melodrama rather than Williams's deeply hued, highly theatrical depiction of the way a small, bigoted town first renders paranoid and then destroys anyone who differs from the norm (later, Williams would refer politely to the production as "under-directed").[3]

He was suffering, too, in his personal life. His lifelong claustrophobia, which had gotten increasingly severe over the years, grew so intense that he almost leapt from a car speeding through the Brooklyn-Battery Tunnel. His drinking habit, which had ratcheted up so mysteriously during the writing of *Cat,* had not abated. By the middle 1950s, he could not walk down a street unless he saw a bar on it, and he was known to flee restaurants if the bar was closed. He would rush home to the security of his own liquor, even if he didn't want a drink. Making matters worse, since the summer of 1955, he had been mixing his liquor with the barbiturate Seconal. His paranoia, which may have been exacerbated by the drinking and drugs, grew, too: increasingly suspicious of friends and unapproachable to strangers, he would stiffen if touched even by an intimate friend. Years later, Maureen Stapleton would relate how Williams's paranoia reached new heights during *Orpheus*'s troubled rehearsal period and brief run. Overhearing negative comments in restaurants, he was convinced they were about him.[4]

Williams's reaction to his father's death was more surprising than his distress over the closing of *Orpheus*. Cornelius and Edwina had separated in 1946, apparently against C. C.'s wishes, ending 37 years of marital strife that wrought severe damage on all the Williamses. While Edwina was already a wealthy woman due to the 50 percent share of the profits of *The Glass Menagerie* that her son had given her, C. C. was surprisingly generous: Although they remained married he gave her their house on Arundel Place in Clayton, Missouri, and half his stock in the International Shoe Company. He returned to his hometown of Knoxville, where he lived briefly with his sister. But their relationship was difficult, as well, and soon C. C. moved into a hotel. A short while later, he met a widow from Cincinnati who became a close companion, drinking and otherwise.[5]

C. C. died in his sleep on March 27, 1957, at age 77 in a Knoxville hospital. Williams and his brother Dakin attended the funeral; Edwina did not. The pressure of these events finally forced Williams to seek the psychiatric help he'd been putting off for three years. And so, in June, Williams paid his first visit to the prominent New York psychiatrist, Lawrence S. Kubie.

II

In the spring of 1957, the attitude of the American psychoanalytic establishment toward homosexuals was not, to put it mildly, constructive. Sigmund Freud, still considered the font of psychoanalytic wisdom, regarded homosexuality as a neurosis, but not as sickness. In

1903, in response to a question posed by the Viennese newspaper, *Die Zeit,* he wrote:

> I am . . . of the firm conviction that homosexuals must not be treated as sick people, for a perverse orientation is far from being a sickness. Would that not oblige us to characterize as sick many great thinkers and scholars of all times, whose perverse orientation we know for a fact and whom we admire precisely because of their mental health? Homosexual persons are not sick.[6]

In 1930, Freud signed a public statement calling for the decriminalization of homosexuality in Germany and Austria. Five years later, in a famous letter written in English to the mother of a homosexual, he elaborated:

> Homosexuality is assuredly no advantage but it is nothing to be ashamed of, no vice, no degradation, it cannot be classified as an illness; we consider it to be a variation of the sexual function produced by a certain arrest of sexual development. Many highly respectable individuals of ancient and modern times have been homosexuals, several of the greatest men among them (Plato, Michelangelo, Leonardo da Vinci, etc.). It is a great injustice to persecute homosexuality as a crime and cruelty too.[7]

Nor did Freud believe that homosexuals ought to be excluded from psychoanalytic training. "We do not on principle want to exclude such persons because we also cannot condone their legal persecution," he wrote to Ernest Jones. "We believe that a decision in such cases should be reserved for an examination of the individual's other qualities."[8]

By 1957, the landscape had changed considerably. With varying degrees of intensity, psychoanalysis was openly hostile toward homosexuality and viewed homosexuals as sick, if not purposefully perverse, individuals. This development, which took greater hold in America than in Europe, was due less to medical or psychoanalytic discoveries than to the politics and culture of post-war America.

During the war, American analysts had held several important posts in military psychiatry. Indeed, one of the most powerful and respected champions of psychiatry in the military was Lawrence Kubie. In 1942, as a member of the National Research Council's Committee on Neuropsychiatry, he formulated the regulation that defined and identified homosexuals for purposes of exclusion from military service. It is most unlikely that Williams knew of Kubie's role in institutionalizing the armed service's prejudice against homosexuals that resulted in their exclusion from serving (and also, if they had avoided detection during the intake process and were arrested while serving, imprisonment, often in mental institutions, and exclusion from benefits under the G. I. Bill).[9]

Following the war, many psychoanalysts became chairmen of the more prestigious departments of psychiatry in American universities. During this period, they worked to transform the still-suspect practice of analysis into the mental health profession of psychiatric medicine. Just as doctors had laid a medical construction on drunkenness (transforming it into alcoholism) and homosexuality half a century earlier, post-war psychoanalysts now fit gayness into a psychiatric disease model.[10]

A genuinely humanistic impulse also contributed to the endemic homophobia of American psychoanalysis by the middle 1950s. Before and during the war, a concerted effort was made by American analysts to rescue their European colleagues from the Third Reich. Hundreds were brought out of Germany, Austria, Hungary, and Czechoslovakia and provided with American sponsors, homes, and jobs. This effort was led by the ubiquitous Lawrence S. Kubie. His courageous crusade, however, would have unforeseen consequences for American gay men and lesbians. In Europe, many of these analysts were radical reformers with Socialist leanings; post-war America was intensely hostile to these aims. The popularity of anti-Communism was not lost on the recently arrived immigrants who, being foreign intellectuals, already felt themselves exposed to possible danger, even deportation. As eager as their American counterparts to transform analysis into a respected and lucrative medical profession, many of the European analysts, soon to form the core of the emerging psychoanalytic establishment, embraced social conformity and orthodoxy, including the public alarm over the supposed menacing homosexual presence in government.[11]

One expression of conformity formulated during this period was the position that being homosexual automatically excluded one from becoming an effective psychiatrist, because the early trauma that "caused" homosexuality produced numerous other "severe personality defects."

Training institutes closed their doors to gay men and women who wanted to pursue a career in psychiatry. However, nowhere did psychoanalysis's conformity to social, cultural, and political norms find as virulent an expression as it did in its response to the findings of Alfred Kinsey.[12]

In a follow-up to his massive report, Kinsey wrote in 1949 that, given the data in his study, "it is difficult to maintain the view that psychosexual relations between individuals of the same sex are rare and therefore abnormal or unnatural," and further that there was no evidence suggesting that "they constitute within themselves evidence of neurosis or psychosis." That people might be troubled by homosexual behavior was due, he wrote, not to psychopathology but to "society's reaction to the individual who departs from the code [of societal norms], or the individual's fear of social reaction." In 1950, a study of collegiate males would show results almost identical to Kinsey's, and no study done in that era would ever claim to refute his findings.[13]

However, not only did the voices of the psychoanalytic establishment proclaim that it had nothing to learn from Kinsey's work, many of the most prominent among them attacked its conclusions on homosexuality with unseemly zeal. The psychoanalyst to speak out against Kinsey with the greatest fervor was Edmund Bergler. Since the early 1930s, as a psychoanalyst in Vienna, Bergler had theorized that homosexuality was a curable disease (but only if the homosexual wanted to change) in which the homosexual suffered from "psychic masochism," the "pleasure of being mistreated," a malady from which every person suffers to some extent. Bergler defined psychic masochism as "the unconscious wish to defeat one's conscious aims, for the purpose of enjoying one's self-made failure." Neurotics exhibit a greater degree of psychic masochism than do normal people; according to Bergler, homosexuals revel in it. The psychic masochism in homosexuals derives from an infant's belief that he had gotten too little from his mother's breast. The infant, however, cannot punish the withholding mother (that would risk death); instead, he turns his anger on himself by depriving himself of the love of women and loving only men instead. This, however, is no more than self-punishment: The homosexual knows, on some level, that his perversion will be punished, not just by society but by other homosexuals, none of whom are happy or healthy, and must penalize others as well as themselves. As Bergler wrote in italics, *"there are no healthy homosexuals"* and he would know: he claimed to have treated almost a thousand. A year before Kinsey published his report, Bergler had written of his gay patients' "megalomaniacal superciliousness" and "their amazing degree of unreliability." "The amount of conflicts, of jealousy, for instance between homosexuals surpasses everything known even in bad heterosexual relationships," he wrote that same year. He marveled at "the great proportion of homosexuals among swindlers, pathologic liars (pseudologues), forgers, lawbreakers of all sorts, drug purveyors, gamblers, spies, pimps, brothel-owners, etc." For years, he made a crusade of refuting the Kinsey Report. Dismissing the findings as "fantastic," he asserted that psychoanalysis "had always considered the homosexual a frightened fugitive from misconceptions he unconsciously builds about women." In 1954, Bergler inveighed against Kinsey in *The New York Post:* "By misinterpreting homosexuality, Kinsey gave homosexuals a new pretext for avoiding medical treatment. He has made it easier for borderline cases, where the issue of homosexuality versus heterosexuality hangs in the balance, to switch to homosexuality." As late as 1959, 11 years after Kinsey had published his findings on male sexuality, Bergler described him as a "voluntary or involuntary dupe of the highly efficient homosexual propaganda machine."[14]

As the president of the New York Psychoanalytic Society, Lawrence Kubie also expressed his opinion of Kinsey's work. Kubie was a relative moderate on homosexuality, and his response to Kinsey was in part a defense of the psychoanalytic point of view. Kinsey, he wrote, suggested that psychoanalysis considered homosexuality ipso facto a sickness, but this wasn't true: "The psychiatrist is far from believing that homosexuality is in and of itself an index of psychopathology . . . but if the analyst's selected experience is in any way characteristic of the whole group . . . the role of unconscious and unattainable goals is greater in the homosexual than in the heterosexual adjustment." But another analyst responding to the Kinsey Report under the pseudonym "Hewitt," voiced, on behalf of his profession, a different view: "Psychoanalysis reveals that all homosexual behavior proceeds as an escape from heterosexual relations based on the fear of such relations. This unequivocally means that all homosexual behavior is abnormal and springs from fear. . . . All homosexuals have severe personality disorders." Kubie and some others voiced legitimate criticisms of Kinsey's methods and analyses, but no one in the profession responded publicly to the distortions of Bergler, "Hewitt," and others.[15]

Tennessee Williams never failed to read up on a subject if it would aid his work. He took the same approach to his analysis: He read several psychoanalytical texts and even took to bandying terms such as "infantile omnipotence" when he was interviewed by Mike Wallace midway through his year with Kubie. If he perused Bergler's 1956 book-length study, published by Hill and Wang, *Homosexuality: Disease or Way of Life?* he would have found this:

> I have no bias against homosexuality . . . [but] homosexuals are essentially disagreeable people . . . [displaying] a mixture of superciliousness, false aggression, and whimpering. . . . [S]ubservient when confronted with a stronger person, merciless when in power, unscrupulous about trampling on a weaker person.[16]

He adds:

> if a homosexual is a great artist, this is so *despite,* and not because of his homosexuality. In the great artist who is a homosexual, a small autarchic corner has been rescued from the holocaust of illness. This corner is constantly invaded and polluted by the homosexual's distorted outlook on life . . . [17]

If Williams had read an account of the 19th International Psychoanalytic Conference, held in Geneva in 1955 he would have learned from the conference's statement on homosexuality that, "Incapacity to love is of course a common characteristic of all the [homosexual] patients we have observed, which confirms, if it be necessary, the immaturity of their personality."[18]

In 1958, the year of *Garden District,* had Williams examined another title from the prolific Bergler, this one, *One Thousand Homosexuals: Conspiracy of Silence, or Curing and Deglamorizing Homosexuals?* he'd have read this warning:

> If information is unavailable, if false statistics are left uncontradicted, if new recruits are not warned by dissemination of the fact that homosexual-

ity is but a disease, the confirmed homosexual is presented with a clear field for his operations—and your teenage children may be the victims.[19]

Perhaps most interesting, Williams would have found, in *Homosexuality: Disease or Way of Life?* Bergler's critique of his novel, *The Roman Spring of Mrs. Stone.* Bergler's discussion is limited to a description of the book's characters, which Bergler catalogues as aggressive lesbians, homosexuals, pimps, and a "urolagnistic-exhibitionist." He summarizes:

> This dazzling lineup of homosexuals, perverts, and bisexuals is a basic part of the narrative; it has not merely been included to confirm Kinsey's opinion that every third man one meets on the street has had some homosexual experience.
>
> This particular framework recalls a statement I made in *The Writer and Psychoanalysis* [1954] to the effect that "writers concern themselves in their work exclusively with abnormal human reactions; normality is not a subject dealt with in poetry of any kind." Mr. Williams seems to overdo this privilege, concentrating as he does so nearly exclusively on perversions.[20]

This was the institution to which Williams presented himself for treatment in the spring of 1957.

III

It may have been a referral from William Inge that, in June, brought Williams to the office of Lawrence S. Kubie. Kubie, who was then 61, had been practicing psychiatry in New York since 1930, and by 1957 was one of the central figures of American psychoanalytic circles. He became president of the New York Psychoanalytic Society in 1939 and held high positions in other medical associations. His list of clients included Inge, Moss Hart, Charles Jackson (author of *The Lost Weekend* and *The Fall of Valor,* a closeted man deeply distressed by his homosexuality), Laura Z. Hobson, and other celebrities.[21] He trained at Harvard, Johns Hopkins, and the National Hospital of Nervous Diseases in London, where his training analyst was Edward Glover, a leading theorist and teacher. He had been a faculty member at Columbia and Yale and on the staff at Mount Sinai Hospital in New York. He wrote prolifically, and by the time he died, in 1977, had produced hundreds of articles and a number of books (including one on creativity, a favorite subject, called *Neurotic Distortion of the Creative Process*). Despite his central position in psychoanalytic circles, Kubie was a gadfly and a challenger of conventional wisdom, more rebel and heretic than orthodox Freudian psychoanalyst. One of the first American analysts of his generation, he would be among the earliest to call for audio and later video records of psychoanalytic sessions and, near the end of his career, for the complete overhaul of the sequence of psychiatric and psychoanalytic training.[22]

Kubie was also a man of tremendous appetites with an enormous capacity for curiosity, learning, and teaching. "He was almost voraciously involved with ideas, people, and activities," one colleague remembered. He published on the psychological effects of hypnotism and, later, of drug therapy; on child psychiatry, sex and marriage, education, war psychology, and the military uses of psychoanalysis; he wrote numerous articles on applied psychoanalysis in art, literature, and religion. One of his interests might have had important resonances in his work with Williams: As one of the relatively small number of psychiatrists in his day who bridged the worlds of psychoanalysis and neuroscience, Kubie was fascinated with the role that the temporal cortex played in memory. On more than one occasion, he recorded the speech of patients being treated with direct electrical stimulation of the temporal cortex.[23]

Kubie's limitless capacity for experience and variety brought to mind, to at least one student and friend, the novelist Thomas Wolfe. Despite his volcanic temperament, students and colleagues recalled him as a gifted teacher and generous counselor. "This famous and uncompromising man was a warm, tolerant, and utterly supportive mentor," remembered one who was both student and later colleague. Kubie could be guilty of what he termed acts of "compulsive benevolence." His leadership in rescuing European analysts from Hitler's concentration camps was only one example.[24]

Kubie's ego was as large as his field of interests, and it revealed itself in several ways, both constructive and less so, from his large humanitarian gestures through the bullying of colleagues (and perhaps patients) and his willingness to be viewed as a maverick. One manifestation was his belief that psychoanalysis was "the most significant event in the recent history of human culture," as well as an almost religious conviction that his chosen profession could, if properly applied, save mankind:

> The patient who comes to analysis refusing to accept anything less than a searching exploration of his personality and development is unwittingly making a contribution to human culture; because the slow accumulation of knowledge which accrues from individual analyses may yet save our civilization and our world, if ever it is used to prevent the development of neurosis in the human infant and child.[25]

While a moderate on homosexuality compared to Bergler and others, and while he argued that the official psychoanalytic stance on homosexuality was more nuanced and flexible than Kinsey had charged and despite his considerable reputation for heterodoxy, Kubie was nonetheless a psychoanalyst in the midst of Cold War culture. His statements on homosexuality per se were comparatively far and few between, but one can glean his feelings on the subject. In *Neurotic Distortion of the Creative Process*, in a demonstration of how unconscious conflicts cripple creativity, Kubie presents an interesting example from all of those he might have chosen:

> A playwright may write half a dozen plays which portray the same theme in various disguises, e.g., a father's struggle to mask his destructive homosexual impulses toward his sons. This theme will be expressed in a series of disguises, without resolution, and with a mounting frustration which colors each successive version. Yet it may remain so well masked that the audiences and critics can no more understand the play than they can understand those neurotic symptoms which assume more banal and everyday forms. Therefore from such art, no matter how artful, no one gains: not the playwright, the players, or the audience. All that triumphs is the impenetrable and insistent rigidity which betrays the failure of art to resolve the neurotic components of the artist and of the culture from which he springs. The neurotic in art is no more self-healing than is the neurotic in the clinic.[26]

Concerning neurotic distortion in the dance, he wrote:

> Insemination rites are danced in which the men are women and the women men. Or if, as sometimes happens, women are allowed to dance the woman's part, they mock the feminine by simpering posture and

gesture as they dance, to the accompaniment of delighted cackles of homosexual laughter in the self-selected audience of balletomanes.[27]

To what degree Kubie explicitly judged his patients' behavior is unknown. The classical analyst maintains a tabula rasa before clients, presenting a blank screen upon which the patient projects his own thoughts
and feelings. While Kubie wrote that this non-judgmental, non-expres-
sive stance was a requirement for any analysis to be successful, there were
also, he believed, exceptions:

> . . . it is traditional and legitimate for most physicians to play a pacifying, re-
> assuring and comforting role toward their patients. The psychoanalyst, on
> the other hand, must often do just the opposite. He will be tactful and judi-
> cious in his warnings, but in the end he must be merciless in forcing a patient
> to face his neurosis. Indeed, just as he must sometimes intervene actively to
> produce situations of deprivation, so he often has to tumble the patient into
> those very situations which arouse his fears, depression and anger.[28]

On at least one occasion, according to Williams, Kubie voiced an unfavorable opinion of his client's plays. It seems likely that on another occasion, they had a discussion about Inge's *The Dark at the Top of the Stairs,*
which had recently opened on Broadway.[29]

According to Williams, Kubie also urged him to give up both homosexuality and writing. At this point, at least according to some of
Williams's friends, Williams stopped taking the doctor seriously: After all,
unlike many gay men who were urged to do so in the 1950s and 1960s,
Williams did not enter psychoanalysis to cure himself of homosexuality.
Kubie may indeed have given such advice; if he did, it was in accordance
with what he described as The Principle of Deprivation: " . . . during his
treatment every patient must be prepared to face periods in which his only
gratification will be a slow and barely perceptible growth in understanding." This requires the patience and ability to tolerate periods of confusion, unhappiness, and deprivation. Sometimes, deprivation will be so
important to a patient's progress that an analyst must intervene in his
daily life and deny him his traditional sources of satisfaction,

> . . . so as to force the patient into a state of active need. . . . For one per-
> son, it may mean living away from home; for another giving up reading or

the movies; another who hides his problems in a compulsive work drive may have to take a vacation from work; a fourth may have to give up a favorite sport or stop even moderate drinking; a fifth may have to avoid normal social life with friends and family. What is cut off will depend on which activity has been used persistently as a major escape from inner problems.[30]

If Williams's reports are to be trusted and Kubie did suggest that he give up both writing and sex with men, these may have been the activities with which, Kubie thought, Williams was avoiding his problems. (Years later, Williams said that Kubie would change the hours of their appointments, making it difficult for Williams to put a time aside for writing. But Williams merely adjusted his schedule accordingly.)

While Kubie may have suggested that Williams give up writing temporarily, such a tactic would have been employed with the intention of freeing Williams from the neuroses that were crippling his creativity. Kubie was fascinated by the creative process, and was a vociferous opponent of the notion, prevalent in the1950s, that an artist's creative roots lay in his unhappiness, his neuroses, and that to create was to effect a self-cure. Merely to be creative, Kubie warned, could not shield an artist from mental illness, nor could it cure him. In his work with creative patients, he set out to prove that the notion that one must be "sick" in order to be creative was a dangerous cliché.[31]

Helping patients gain insight to the conflicts generated in their unconscious was the center of his practice. For Kubie, the unconscious was a tyrant whose desires and drives remained unknown to the patient, and as long as they were veiled in mystery, the ill person could not heal his life, nor could the artist be truly free to create. Behavior could be called abnormal as long as it was primarily determined by factors outside of the patient's awareness and unavailable for his examination. In Kubie's psychoanalytic topography, the key to creativity lay not in the unconscious but in the *pre*conscious. This was that process between the limited, language-oriented conscious mind, and the unrulable unconscious. Relying not on language but on the swift processing of abstract concepts, the preconscious makes connections between symbols at a much faster rate and with much greater flexibility than does the relatively slow, literal, conscious mind. Symbols in the preconscious overlap and express many more meanings than can be expressed in a few conscious words. In an example Kubie used, the abstract concept of "chair," on the preconscious level, sets off many reverberations "down many mental corridors, all of which are tagged by the coded symbol 'chair.'" Clear communication in daily speech, however, requires that we assign images only one meaning at a time. Thus, we do not think of every one of a word's possible images when we use the word in conversation. The conscious mind samples the preconscious for the specific instance and image of "chair" it requires to make its point with clarity. "How much do you want for that chair?" is understood by both parties to refer solely to the object under discussion, rather than to any other idea and instance of "chair" in their respective experiences. Meanwhile, the manifest other images and meanings of "chair" remain active on the fringes of our conscious mind, like, Kubie wrote, "the sound of distant music." Such preconscious echoes or condensations of multiple meanings and images are used in poetry, humor, the dream, and the symptom. They are also the basic ingredients of art. In theory, we all have access to these newly created images. The artist, according to Kubie, is one who, through some mysterious accident, has retained the ability to use his or her preconscious faculties more freely than other people who may be potentially as gifted.[32]

In order for the preconscious to make a significant contribution to an individual's creativity, it must be free to gather, assemble, and rearrange ideas and images. An unfettered preconscious supplies the artist with a constant stream of old data—images, concepts, and information—rearranged into new combinations for the artist's use. In other words, the preconscious reorganizes the material of everyday life into images an artist can employ. However, the preconscious is hampered in its operation on the one hand by the rigidity of the plodding conscious mind's everyday limitations of precise, literal language, and on the other, by the "unreality" of the unconscious mind's fears, aims, and impulses that are outside the corrective influence of experience, and which are rendered impenetrable through distortion and disguise. Therefore, the forces that originate primarily in the unconscious, such as guilt, shame, rage, and hatred, and the rigid patterns of behavior they foist on the unknowing artist, limit the free-ranging creative span of her or his preconscious. In this circumstance, old data cannot be rearranged into new patterns and meanings. Only by becoming aware of those unconscious conflicts can the artist restore to his or her preconscious its ability to gather, sort, and recombine its symbols freely.[33]

Kubie's own mind worked best in such lightning flashes, in the creative rearrangement of the material at hand. When his former student and long-time colleague Eugene B. Brody was exploring the uses of making audio recordings of analysis sessions, Kubie acted as a consultant, listening to the tapes and commenting. His remarks had flashes of unconventional brilliance about them; Brody would recall that their bases lay in Kubie's intuitions, which could be startlingly good, but highly idiosyncratic. Kubie possessed a creative temperament as well as an artist's ego, which he demonstrated in his willingness to recombine elements of Freudian conventional wisdom into new methods of treatment and education. Kubie's artistic temperament may also have extended to his demeanor with his more artistic clients. He tended to talk more than most of his analyst colleagues, and he could be unusually competitive with his more illustrious patients. Apparently, on occasion he even introduced them to each other: The playwright and screenwriter Arthur Laurents claims that Kubie introduced Moss Hart to another patient, Kurt Weill, and the result was their musical *Lady in the Dark*. Then, under a pseudonym, Kubie wrote an Introduction to the published text. Indeed, it is possible that one of his primary interests coincided with one of Williams's, and may well have aided in the conception of Williams's next major work.[34]

During the course of his analysis, Williams must have talked at length about his sister, her illness, her lobotomy, his feelings about Edwina's and C. C.'s roles in it, and his own guilt for surviving while Rose, in a crucial sense, did not. If Kubie were indeed unusually talkative and competitive, he might have mentioned his own abiding interest in the temporal lobe, and his taking down the words of a patient, unconscious on an operating table, whose temporal lobe was being stimulated. This, in turn, may have stimulated Williams's imagination. Kubie was particularly interested in the role the temporal lobe played in memory, in integrating past and present events, and thus in the creation of personality. The temporal lobe, he wrote, was where "the 'I' and the 'non-I' pole of the symbol meet." The integrating function of the temporal lobe is precisely what

Violet Venable wants to interrupt in the brain of her niece in *Suddenly Last Summer;* what Catharine fears the most is the erasing of her memories, the disintegration of past events and the destruction of her character.[35]

Williams, of course, did not need the professional history of Lawrence Kubie to be interested in the effects of the surgical destruction of the temporal lobe. It may be coincidence that the play he wrote in the mornings while seeing Kubie later in the day had, at the center of its action, the spectre of an operation in which both he and his psychiatrist shared a special interest. It may also be a coincidence that the character in *Suddenly Last Summer* who holds the central figure's fate in his hands practices neurosurgery, a specialty in which Williams's psychiatrist was, for his time, unusually well-versed. It may be a coincidence that this doctor has the peculiar name *Cukrowicz,* which, the character tells us, is Polish for sugar. Since Violet has trouble pronouncing it, the good doctor simplifies it: "Call me Dr. Sugar," he says. From Catharine's point of view, it is this doctor, Dr. Sugar/Kubie, who stands between her and the death of her personality desired by the fearsome mother-figure Aunt Violet, whose relationship with her son had about it more than a few overtones of incest. Not surprisingly, the end of the play leaves Catharine's fate at the hands of Dr. Sugar/Kubie unresolved.[36]

On retirement, Kubie destroyed his patient records, so what he might have revealed about his own life or opinions in the course of Williams's analysis, including any mention of his interest in the temporal lobe and its connection to memory, is a matter of speculation. Nonetheless, however unusual it might have been for a Freudian analyst of the 1950s to offer examples from his own life or personal opinion of a patient's work, Williams wrote to his mother that Kubie had done just that: "He hit me where it hurt most," he wrote Edwina. "He said I wrote cheap melodramas and nothing else." If Kubie did in fact voice such a personal, damaging opinion (as much an exercise in ego as in criticism), it is at least possible that he might also have discussed his own scientific work.[37]

Williams was as forthcoming in the press about his psychoanalysis as a child displaying a new toy. "I think if this analysis works, it will open some doors for me," he told the *New York Herald Tribune* six months into his sessions with Kubie. "If I am no longer disturbed myself, I will deal less with disturbed people and violent material. . . . It would be good if I could write with serenity."[38]

He also reported that his sessions with Kubie were affording him "the most enormous relief." At the minimum, his year with Kubie substantially reduced his claustrophobia: he was able to ride without panic in the tiny one-passenger elevator in his apartment on the East Side. The sessions also hastened his coming to terms with his father, a process that seems to have begun before C. C.'s death. As late as 1970, he told an interviewer that he considered his sessions with Kubie a success. To be a patient of Kubie's, however, meant committing to five daily sessions a week, and that, in turn, meant staying in New York for long periods of time. Williams could rarely stay anywhere for any length of time, and once he had wealth, he almost never spent more than a few weeks in any one place, from New Orleans to Key West to Taormino to New York, a city he never much liked. He told the *Herald Tribune* that he didn't intend to become a permanent analysand, and planned to give it up after a year.[39]

Soon enough, his restlessness overtook him and he began planning his escape.

> I'm getting away again, Sunday, to Florida, [he wrote Edwina and Dakin] as I felt myself reaching the point of exhaustion. Dr. _____ opposes the move but I think I have to consider my physical state as well as what he thinks is of psychological value, i.e., staying with him in New York. I respect the doctor and feel he's done me some good but his fees are too high and if I continue analysis next Fall, it will probably be with someone younger and less expensive to go to. Or maybe I won't feel the need of continuing it at all.[40]

Over time, he gave several reasons for breaking with Kubie: he could no longer take the doctor's advice seriously; his fees were too high ($50 an hour); Kubie's volcanic nature reminded him too much of his father and frightened him. He ended his treatment after a year, although he continued a part-time flirtation with analysis for the rest of his life. (He spent part of the summer he quit Kubie at Austin Riggs, an expensive private mental hospital in Stockbridge, Massachusetts. William Inge had done a stint there, and, as he may have referred Williams to Kubie, he may have recommended Austin Riggs, as well. Perhaps not coincidentally, Kubie played a role in the early years of that institution, serving on its Medical Advisory Board and Board of Trustees.[41])

The truth is that Williams was probably not a good candidate for analysis, at least not as it was practiced by Freudians (even mavericks like Kubie) in the 1950s. Kubie saw the analyst's job as strengthening "[an] individual to the point at which he will be able to face and accept the whole truth." In theory, this ought to have made him a good match for Williams who, after all, in interview after interview and play after play proclaimed his devotion to the truth. As he told the *New York Herald Tribune* while *Orpheus Descending* was in rehearsal, "The moral contribution of my plays is that they expose what I consider to be untrue." In the play Williams wrote while in analysis, Catharine would proclaim, "The truth's the one thing I have never resisted!"[42] (401).

Williams's devotion to the truth, however, was tempered by his even deeper commitment to freedom. "I want to be free and have freedom all around me. I don't want anything tight or limiting or strained," he wrote as a young man. The need for freedom and fear of confinement never left him. Freedom, Kubie would say, was exactly what he offered Williams, but his definition differed significantly from his client's. To Kubie, freedom meant relief from the neuroses that, in his view, crippled Williams's writing and his life. (Williams must initially have thought so, too, or he wouldn't have sought an analyst.) Kubie was a steadfast soldier in the battle against repression and in favor of self-knowledge: He agreed with Freud that repression is an attempt at flight from meaning. To Williams, freedom meant flight from the sources of his conflicts, either physically or in the incomplete release and solace he found in constructing symbols from them in plays, stories, and poems. Williams's instinct was strongly, but not entirely, for flight (and in this sense, for repression). The battle between concealing and revealing that Williams fought out in his work, his instinct for truth versus his self-characterized "gift for evasion," drove him first to analysis and then from it. Requiring as it did a relentless examination of the conflicts in his life, with the oversized and frightening personality of Kubie as inquisitor, the freedom offered by psychoanalysis came at too high a price. In the end, it was more than Williams was will-ing to tolerate.[43]

Williams's writing, especially a piece like *Suddenly Last Summer,* composed in the throes of analysis, is both an engagement with and rev-elation of violent feelings and wishes, and an evasion and concealment of them. By casting them in the form of fiction, attaching them to fic-tional characters and detaching them from himself, Williams was able to put these feelings and wishes at a distance that made them amenable to his artistic control. In this indirect way, he exorcised these feelings and relieved himself of their psychic pressure. (That this was precisely what Rose had been unable to do added to his lifelong burden of guilt, and brought extra urgency to the conflicts of *Suddenly Last Summer.*) That Williams was able to deploy such charged material in so efficient and ef-fective a way—and that he would never command such tight control again—suggests that psychoanalysis may well have been helpful to him, or at least to his work. Analysis certainly didn't ameliorate the destruc-tive way in which he led his life; nor did it seem to have any long-term effect on his writing. Analysis didn't grant him his wish of writing less violent material about disturbed people: He followed *Suddenly Last Summer* with *Sweet Bird of Youth.* But it may be that by forcing him to engage the fearful passions of his inner life, his daily meetings with Kubie relieved him sufficiently so that those violent feelings did not overwhelm his work. There is no guarantee, of course, that analysis would have continued to provide him with just enough relief to use his emotions so effectively in his writing. It would prove unfortunate that Williams's need for escape made it all but impossible, over the long run, to siphon off neurotic pressures through either creation or psychother-apy. Alcohol and drugs provided the flight from work, and work, in the end, was his only salvation.

IV

In June 1957, just after *Orpheus Descending* closed and at about the time he began seeing Kubie, Williams took a vacation to Havana. Most likely while staying at the Hotel Comodoro, Williams started a one-act called *And Tell Sad Tales of the Deaths of Queens.* . . . It is a play about the things one does—sometimes foolish, desperate things—to stave off the spectre of loneliness.[44]

The Queen, in this instance, is Candy, a newly single interior designer who is about to turn 35. Williams's description of Candy is instructive: "*The effeminacy of Candy is too natural, too innate, to require expression in mannerisms or voice: the part should be played without caricature.*" For any remarks Williams may have made about disliking effeminate men and drag queens, here he instructs actors and directors to play one with naturalness and simplicity.

For 17 years, Candy had an older lover who had set him up in business, but his "sponsor" has recently deserted him for another younger man. Candy is now the sole owner of the interior design business, and of three pieces of lucrative rental property in the French Quarter in New Orleans. Prey to the emotional and physical violence often visited on gay men, he has managed to construct his own world, one that keeps him safe (usually) from the unfriendly one outside. Loneliness will find Candy even here, however. Now that Sidney Korngold, the former lover, has left, Candy is forced to turn to the world outside for companionship and love, and he has invited into his home a brutish, rather dim sailor named Karl, whom he met earlier in the evening at a gay bar in the French Quarter. So afraid of encroaching loneliness, Candy invents a love relationship with Karl. Needless to say, it will not end well.[45]

The world that Candy has made for himself is thoroughly gay: his clients are gay, his tenants, including the two young men who live upstairs, are gay. Straight people appear very rarely in this world. He tells Karl,

> I'd never consider renting to anyone else [besides gay men]. Queens make wonderful tenants, take excellent care of the place, sometimes improve it for you. They are great home-lovers and have creative ideas. They set the styles and create the taste for the country. Don't you know that? [. . .] Just imagine this country without queens in it. It would be absolutely barbaric. Look at the homes of normal married couples. No originality; modern mixed with period, everything bunched around a big TV set in the parlor. Mediocrity is the passion among them. Conformity. Convention. (396)

The speech brings to mind Tom Wingfield's description of his apartment building in St. Louis and his desire to escape that world into one in which he can be his own authentic self, sexually and otherwise.

Tucked within this safe haven is another: a Japanese garden that Candy has created on the patio. It contains a pool spanned by a footbridge, an eighteenth-century bench, and a willow tree. It is meant, one feels, to be a refuge, as different from the adjacent French Quarter as it could be, a place where Candy can come to be safe from the "squares" Williams wrote about so often (such as the Matron and her drunken husband in *Lord Byron's Love Letter*). There is yet another hideaway within the Japanese garden: "In the middle of my fishpool is an island with a willow that makes a complete curtain, an absolutely private retreat from the world except for a few little glimpses of the sky now and then . . ." (398). Like Lady's confectionery in *Orpheus Descending*, the hideaway provides a fantasy of shelter more than any real defense from a world whose constants are rejection and loneliness.

Not all is entirely well within this little world. In addition to the loneliness that has insinuated itself amid all the *japanoiserie*, there are the two disruptive gay tenants who live upstairs. While Candy at first refers to them as "a pair of sweet boys from Alabama," (396) it becomes clear that he doesn't approve of their nocturnal habits. Already deep into a romantic fantasy, Candy allows himself to think that what a sailor like Karl is looking for in a man is respectability, and when Alvin, one of the tenants, comes calling while Candy is entertaining Karl, Candy attacks: "You cruise all night and bring home trick after trick which I put up with despite the chance I'm taking of making a terrible scandal" (407). Candy himself, it seems, has always been monogamous, even while Korngold was not; further, Karl is the first man Candy has brought home since Korngold left. More abuse of the tenants follows. When the other tenant, Jerry, comes calling to present Candy with a 35th birthday gift, Candy calls him and Alvin "bitches" and demands that they move out.

It is easy enough to dismiss this play as more self-loathing claptrap: The leading gay character is an effeminate man who talks himself into believing he loves a total, and unkind, stranger. Candy is generous to a fault and, after having spent $300 dollars on the sailor, Karl beats and robs him for his trouble. Moreover, Candy is cruel to his young gay tenants, insisting in no uncertain terms that they stop camping around like tramps. This is Candy's fastidiousness, however, not Williams's homophobia. It is part and parcel of the way Candy tries to control his environment—like the way he decorates his home and creates a romantic fiction in his mind. For their parts, the reason Alvin and Jerry come downstairs in the first place is to make sure that Candy is all right: They've heard about Karl, and know how he broke the jaw of a friend of theirs. Had audiences seen this play in 1957, they would have witnessed something new to them: gay men looking out for one another.

Further, if Karl is the brute and Candy his victim, Williams deepens and complicates the relationship by making Candy the stronger of the two—he knows who he is, which is more than can be said for Karl. The sailor proclaims surprise when he learns that Candy is gay, and insists that he doesn't "go with queers" (397). He forgets the fact that they met in a gay bar, and that he has been hanging around such places for years whenever he's in port.

When Candy displays himself in drag, Karl is not entirely certain he knows who *she* is, either, "You're as much like a woman as any real one I seen," he tells him. (401) He asks if Candy's certain he's not one, and Candy offers to show him—an offer Karl declines in a single abrupt syllable. With some drink in him, perhaps Karl decided he was hungry enough for sex that he would go home with a man. Now, however, in the gay man's world, he is uncertain. He changes his mind about sex (at least for the moment) but can't summon the willpower to leave and find a woman. When Candy asks him to dance he refuses. Only when she challenges him—"Are you afraid to?"—does he agree. (403) Even then, he can only do so for a moment before quitting. "I can't," he says. "It seems too [. . .] unnatural—not right. I'd better go" (403). Candy challenges him again: "Don't be so conventional and inhibited, why, what for!?" (404). In her world, at least for the moment, she is the one who pushes and prods, the aggressor challenging the square to break out of prejudice and habit. Karl has brute force, but Candy, in his need for love, can stand up to it.

Karl can't break out of narrow-mindedness, but he remains ambivalent about leaving. He demands that Candy find him a "real" woman to have sex with—which Candy does, after paying Karl to simply provide him with companionship. After his straight assignation, Karl returns, but only, he insists, because he's broke. He says his "real" home in New Orleans is a bed at the Salvation Army, when he has the money to pay for it.

Candy may live in something of a dream world—the apartment's decoration and his belief that he loves Karl certainly suggests so—but his fantasies are a fortress considerably less central to his survival than the one Karl has built. Candy doesn't lie about his identity; Karl clings to the tiniest shreds of evidence to conceal from himself the true nature of his longing. He's even willing to let Candy fellate him—for ten dollars. The monetary transaction will allow him to still think of himself as manly and Candy as a prostitute. To Candy, however, the money represents only one way of exhibiting generosity and friendship; it doesn't change the nature of their relationship as he imagines it.

In the end, Candy debases himself before Karl by insisting that he loves him, even though, feeling so threatened by the unconventional life he offers him, Karl beats him severely and steals his money. Williams, too, had been beaten by rough trade in the St. George Hotel in 1943, and just as his biggest fear, then and later, was loneliness (*"Give me time for tenderness,"* he wrote in his journal afterward), so is Candy's. When Karl, unnerved by dancing with a man, says he's going to leave, Candy's reaction is both pleading and aggressive: *"OH, NO!—NO!!"* (403). Candy survives, however, and, with the aid of his two young tenants, even laughs at his plight. One senses he will soon be back in the bars, looking for love, even alongside Alvin and Jerry.[46]

And Tell Sad Stories of the Death of Queens is a minor effort compared to the next play that Williams would write while in analysis with Kubie. Still, a play written in 1957 that shows an effeminate gay man as psychologically stronger than a straight man can't be altogether dismissed. What cripples Candy—what causes him to humiliate himself before Karl—is loneliness. Denial of one's own true sexual feelings is the straight man's game.[47]

V

One result of Williams's year with Kubie was a deepened understanding of his father. While his analysis aided his reaching a posthumous rapprochement with the turbulent C. C., the process itself seemed to have been underway before Williams met Kubie. In *Cat on a Hot Tin Roof,* Williams created a father whose larger-than-life personality was not only not frightening to his favorite son, Brick, but was, in its rough way, welcoming and comforting. Big Daddy reaches out to the suffering Brick, and offers to wrap him in a masculine cocoon of love and protection. Here was a father who loved his son and could show it. Moreover, Big Daddy preferred that son to the other, more conventionally successful Gooper. For Williams, this was a consummation devoutly to be wished.[48]

Early in Williams's life, his father was first the man whose occasional stops at home were terrifying episodes of harsh language and violent behavior. Later, in St. Louis, C. C. would enter the house each evening "as though he were entering it with the intention of tearing it down from the inside" and proceed to his alcove bedroom, down the hallway from Edwina's, where he would close the door and drink. By the time Williams wrote the prose tribute to his father called "The Man in the Overstuffed Chair," about 1960, he recognized Cornelius as "the Mississippi drummer, who was removed from the wild and free road and put behind a desk like a jungle animal put in a cage in a zoo." Williams had come to recognize his own sometimes overwhelming physical and spiritual claustrophobia as his father's ailment, as well.[49]

What Williams had realized about his father was C. C.'s deep, unexpressed feelings, his alienation from his wife, his wife's family, and his two eldest children (the only child he was close to was the youngest, Dakin). Williams also may have suffered considerable guilt about absorbing from his mother such an intense dislike for Cornelius whom, he learned too late, he had misunderstood as completely as his father had underestimated him. While C. C. despised the "Nancy boy" he felt Edwina had made of Tom, seeing Tom as a rival for her affection might also have fueled his dislike for his son. All of Edwina's love for any male of her household went to Tom, then to Dakin, never to her husband.

Williams wrote of his usually placid childhood days, "Only on those occasional weekends when my father visited the house were things different. Then the spell of perfect peace was broken. A loud voice was heard, and heavy footsteps. Doors were slammed. Furniture was kicked and banged . . ." Edwina quotes this passage in her book, *Remember Me to Tom,* and she was pleased to inculcate in her children this picture of Cornelius as irrational, vulgar, and brutish. But was Cornelius, alone, responsible for scenes like this? Did Edwina, whose disdain for her husband seemed to begin almost as soon as they were married, who locked him out of her bedroom and only occasionally admitted him, usually with protests and screams, play no part in their creation at all?[50]

It may be that a fuller, fairer picture of his parents' unfortunate marriage was on Williams's mind when he wrote *Something Unspoken* in 1953. It was not produced until 1958 when, paired with *Suddenly Last Summer,* it premiered under the umbrella title *Garden District.* Since their premieres (when *Something Unspoken* rated barely a mention in reviews), *Suddenly Last Summer* has garnered almost all the attention. Given its melodramatic content, its sharp and mordant conflict, and atmosphere of barely suppressed hysteria, this is not surprising. *Suddenly Last Summer* is constantly revived; *Something Unspoken* hardly at all. Why did Williams choose this small, modest, seemingly trivial piece to pair with a wild sister so likely to devour it?

If the popular interpretation of *Suddenly Last Summer* is to see it as a condemnation of Edwina's treatment of Rose Williams, *Something Unspoken* can be seen as reflections on at least two aspects of the Williamses' St. Louis home life: the relationship between Edwina and her husband, and that between Tom Williams and his father.

On the surface, however, the play demands attention as Williams's one published play whose principal characters are lesbians. They are the imperious society matron Cornelia Scott and her faded, mousy secretary, Grace Lancaster. The play takes place in the parlor of Cornelia's house in the town of Meridian, on the day when the Confederate Daughters are having their annual election. It is also the fifteenth anniversary of Grace's arrival on Cornelia's doorstep. Cornelia is embroiled in an attempt to be elected Regent of the Daughters by acclamation, all the while insisting to Grace that she has no interest in the post. Refusing to go the meeting,

Cornelia keeps in phone contact with her lieutenant, Esmeralda Hawkins, who is supposed to be arranging Esmeralda's ascension in the teeth of an opposing faction of nouveau members: women who could not "get a front pew at the Second Baptist Church!"[51]

On this anniversary day, Cornelia is also trying to get the attention of her secretary: she leaves a single rose in a crystal vase on the parlor's mahogany table and 14 more on Grace's desk in the library. But Grace is unwilling to acknowledge what their relationship means to Cornelia. The play's tensions are the result of the awaited outcome of the election and the conflict between Cornelia and Grace over what will and what will not remain unspoken between them.

If both characters are lesbians or at least have powerful lesbian feelings, those feelings have never been consummated. Rather, power is the fulcrum in this relationship. Cornelia—imposing, Romanesque, holder of many offices in many Meridian women's associations (but never, as she points out, regent of the Confederate Daughters)—would seem to be the dominant partner, while the faded, retiring Grace remains the junior. It is Cornelia who insists on speaking what is unspoken between them, while Grace is made so anxious by the possibility that she tries to leave the room whenever her employer brings the subject up. But Williams quickly undermines the first impression of who wields the power in this relationship. Grace holds Cornelia in a certain amount of contempt and must work hard to conceal it. When Cornelia describes how she has rallied her forces within the Confederate Daughters to defeat the cabal aligned against her, Grace's lips *twitch slightly as if she had an hysterical impulse to smile.* "Your—*forces?*" she says (281). Grace, however, also works hard at portraying Cornelia as the stronger one and herself as the weaker. "You're the strong one of us two and surely you know it," she insists (292). Cornelia is easily understood: she wants love, and she wants it unconditionally, in her public life by acclamation of the Confederate Daughters, and, intimately, from Grace. Grace is opaque and even devious; her agenda has to do with control. To withhold from Cornelia a statement of her love, to forbid Cornelia from speaking of her own, is to leave the older woman in a position of weakness, of always needing to know and never knowing. As long as the something unsaid remains unspoken, Grace is the dominant partner. Once any feelings of love are spoken, once Cornelia is allowed to make her declaration and Grace reciprocates, then they are equals. Grace is unwilling for that to happen and Cornelia is reluctant to force the issue, perhaps out of fear of losing Grace.

It becomes equally clear that Cornelia is weak. She wants the approval of the Confederate Daughters but cannot admit to Grace she wants it, cannot campaign for it nor even allow herself to be chosen from a pool of nominees: the Daughters' approval must be unanimous. Nor can she have her way with Grace. In the end, she is defeated on both fronts: a mediocre parvenu is elected regent, and Grace succeeds in keeping any expression of love unspoken. As she leaves the room to fetch paper and pen to take Cornelia's letter of resignation from the Daughters, the stage direction reads, *she turns to glance at Cornelia's rigid shoulders and a slight, equivocal smile momentarily appears on her face; not quite malicious but not really sympathetic* (296). "All my life," Williams wrote in *The New York Times* a year after *Garden District* opened, "I have been haunted by the obsession that to desire a thing or to love a thing intensely is to place yourself in a vulnerable position, to be a possible, if not a probable, loser of what you most want." Cornelia makes herself vulnerable and loses the position she desires both in society and in her own house; Grace refuses to do so, and retains her power over her employer.[52]

What Cornelia wants most, what she needs to surround herself with, within her home and without, and what finally are denied her, are words of love. Her extensive record collection is filled with recordings of lieder and she orders more from the Gramophone Shoppe in Atlanta: The house must be filled not only with music but with musical expressions of love.

However, while Grace may have the upper hand in this silent battle of wills, she is also threatened by certain limitations. Like Cornelia, she is a spinster, but Cornelia is a woman of independent means, while Grace, at the age of 45, relies on the love and admiration of her employer for whatever position she holds in the community. Should Grace relinquish her power over Cornelia (and possibly over her own emotions, as well) by allowing their feelings to be spoken and perhaps acted on, she risks upsetting the status quo. In a house ruled by silence, Cornelia must live in hope: Perhaps tomorrow Grace will give way and the magical words, "I love you," will be spoken, and then, perhaps, in some fashion acted on. If Grace in fact loves Cornelia, she is too frightened of such a terrible, unconventional emotion to say so. If she does not love Cornelia and Cornelia discovers this, then her position in Cornelia's home is threatened, and she risks being asked to leave. Where will a 45-year-old spinster go? In the world of Tennessee Williams, being thrown out of one's home is the direst nightmare, and may lead, in the end, to the madhouse.

Cornelia's desperate longing for love and Grace's equally strong need for a safe haven went unnoticed by *Garden District*'s critics. Even the more positive notices passed over it quickly. "The first, 'Something Unspoken,' is the better of the two," said the critic from *Cue*, "dissects [*sic*] the rather complex, ambivalent relationships between two elderly ladies." *Theatre Arts* similarly allowed the play one sentence in its column-long notice. Henry Hewes mentions it not at all in *The Saturday Review;* Brooks Atkinson considered it a "trifling, inconclusive anecdote" and dealt with it in one paragraph at the bottom of his review.[53]

The following September, Kenneth Tynan gave the London production two paragraphs in his *Observer* review, and the perceptive critic may well have identified the coup de grâce delivered to the play: the direction by Herbert Machiz (who directed *Garden District* in New York, as well). Machiz, according to Tynan, staged the play as a farce: "It is like seeing vintage Strindberg performed by Punch and Judy," he wrote. *Something Unspoken* might not compete with the assault on a late-1950s audience's sensibilities launched by *Suddenly Last Summer,* but Tynan was not far wrong when he compared it to a piece by Strindberg.[54]

It is beside the point, even if it were possible, to make direct, one-to-one correspondences between the events of an artist's childhood and the themes and plots of his work (and one will find very few such connections in *Something Unspoken*). What is useful, however, is to discover the emotional ties that bind an artist's life to his art, in order to stage his plays with greater awareness of the underlying motivations and circumstances that lead to action. If, for example, one imagined the stifled emotions and tensions, the years of unspoken words in any of the Williams's St. Louis homes and the wish, especially of the Williams children, for an end to the tension and the speaking of words that needed to be said; if one considered the upper hand held by the diminutive Edwina, who could defeat her large and fearsome husband in any verbal confrontation and who withheld from him every expression of emotional and physical love; and if one could imagine, as Williams came to do, that deep within his volcanic, furious nature, Cornelius Coffin Williams—deprived of a mother from the age of five and shunted from one aunt to another before finally being sent to a seminary and then military school, unable to express deep emotions—desperately needed someone to say that she loved him, then one can begin to understand that *Something Unspoken* is not a silly little play of minor characters wanting small things. Like *The Glass Menagerie* and Williams's other best work, *Something Unspoken* is a play in which the characters want large things desperately. Should a director and two actors find the depth of desperation, then *Something Unspoken* might indeed work, even as a curtain-raiser to *Suddenly Last Summer.*

Something Unspoken takes place on the fifteenth anniversary of Grace's employment. When Williams was 15 years old, C. C. moved the family from a small apartment in what was, in Edwina's eyes, a reasonably fashionable section of St. Louis on Cates Street, to one on Enright Avenue in University City, just over the city line. The move was made in order for Tom to enter University City High School. The new apartment was even smaller than the one on Cates Street and the tensions were exacerbated between Cornelius and Edwina, between Edwina and Rose, and between Cornelius and Rose and Tom. There was no escaping the emotional unhappiness in this dismally overcrowded place, and Williams came to hate the close, claustrophobia-inducing apartment (but he did not forget it: it was the inspiration for the apartment in which he set *The Glass Menagerie*).

The atmosphere in the Williams household, especially once the family had relocated to St. Louis, was almost always suffocating and tense. In accordance with the times, C. C. considered himself the rightful master of his home, but Edwina would not be mastered. Cornelius's deepening, and finally bottomless frustrations are not so different from Cornelia Scott's inability to be mistress over her secretary, although Grace rules not by chatter, as Edwina did, but by silence.

The Williams household could not be characterized as one in which most words went unspoken—no house where Edwina lived would ever be silent, as she emitted an unending stream of relentless conversation as naturally as she breathed, and Cornelius's angry outbursts would be punctuated by the slamming of doors and furniture. There were arguments about money, about C. C.'s drinking, about Edwina's supposed spoiling of the children. Williams's lifelong need for escape, even, perhaps, his claustrophobia, can quite possibly be traced to the evenings when Cornelius and Edwina were both at home—although they took pains to occupy separate rooms whenever possible. When Edwina's parents came for extended visits, the atmosphere was even worse. While Cornelius treated Edwina's mother in a gentlemanly fashion, he never hid his contempt for her minister father.

At the age of 16, after a year in the smothering confines of Enright Avenue, Williams published his first significant literary effort in a national magazine. He entered a contest sponsored by *Smart Set* magazine, writing a brief essay in answer to the question, "Can a Good Wife Be a Good Sport?" Assuming the role of a world-weary husband, Williams describes life with an uncontrollable woman who does exactly what she wishes, when she wishes. The unnamed husband describes his wife's drinking, her smoking, her sexual affairs, her "staying up all night with the boys in cabarets." Eventually, the wife obtains a divorce, leaving her ex-husband to tell his cautionary tale. Williams's essay won him third prize (among the 12 winners, he was the only male), but the award he might have been seeking unconsciously was his father's approval. He had, after all, told the story of a demoralized, humiliated husband from the man's point of view with sympathy and, for a 16-year-old, acute insight. As literature, the piece is sophisticated. Williams gives all of Cornelius's bad habits to the wife, so that none of the family, or any who knew them, could say that the piece was based on the teenager's own home life. (Edwina, in her autobiography, proudly reprints the piece in full, never offering a suggestion that she had the slightest clue as to where her son's inspiration might have come from.) In *Something Unspoken*, the grown Williams does something similar, dividing and recombining aspects of Edwina's and Cornelius's personalities between Grace and Cornelia. Cornelia, who so needs to be told that she is loved, puts great emphasis on her social position (just as Edwina was regent of her Daughters of the American Revolution chapter, and in *Remember Me to Tom* proudly describes not only the Dakin coat-of-arms, but that of her husband's family, as well). Both

Cornelia and Grace will chatter to cover up embarrassing silences, but the diminutive Grace uses silence as a weapon to defeat the larger, gregarious Cornelia.[55]

In *Something Unspoken,* Grace, defending her need for silence, says to Cornelia, "To speak out things that are fifteen years unspoken!—That long a time can make a silence a wall that nothing less than dynamite could break through [. . .]" (292–3). This wall also existed between Cornelius and each member of his St. Louis family excepting Dakin. Williams included himself in his wish to tear down the wall of silence; the unspoken love in *Something Unspoken* is also that of the playwright for his father.

As different as *Something Unspoken* and *Suddenly Last Summer* are in emotional pitch and materials, they are in other ways quite complementary. Both rely on Williams's usual tension between the need to conceal and the urge to reveal: While Cornelia wants to reveal the feelings she and Grace have for each other, Grace is intent on concealing them; Catharine wants to reveal the truth about what happened to Sebastian at Cabeza del Lobo, while her aunt Violet must silence her. Violet will go to any length, including lobotomizing her niece, to deny the truth about Sebastian; and Grace must also deny the truth of the emotional ties between her and Cornelia. Catharine famously reflects on the uses we make of each other while calling it love, while Grace is quite willing to use Cornelia in order to retain her position. The thing that is absent plays a large role in both pieces: Sebastian is physically absent, although his presence hovers, while the facts of his life—that arrangement of words and images that will replace his physical being—are desperately fought over. Similarly, Cornelia and Grace silently wrestle with the words that have been absent in their relationship, which would give it a specific meaning. In this sense, the two plays, seemingly so different, are concerned with the same thing: finding or suppressing the words that will give specific meaning to a relationship unsanctioned by society.

VI

What is to be done with *Suddenly Last Summer?* For critics or directors interested in Williams's images of homosexuality, the play is what *The*

Merchant of Venice is to Jewish critics of Shakespeare, or *The Taming of the Shrew* to feminist Shakespeareans. At least the "offensive" characters in these other plays actually appear onstage and may be able, with directorial intervention, to make a case for themselves. If Shakespeare's portrayal of Shylock began as either a stereotypical Jew or, as may be equally likely, the standard *commedia* Pantalone figure, the playwright's artistry got the better of him. If the same cannot be said of his treatment of Kate in *Taming of the Shrew*, at least directors can invent stage business in the last scene of the play that ameliorates what looks like misogyny (even if such business almost always seems a tacked-on, inorganic apology by an embarrassed director). Williams, however, seems to give sympathetic directors and critics no such chance. Sebastian Venable is offstage and dead and there is no denying his predatory nature or his sickening end. Unlike the spiritually paralyzed Brick or supremely neurasthenic Elois Duvenet, Sebastian is a predatory pedophile who is literally devoured by a rampaging pack of his victims.

To Nicholas de Jongh, the play is beyond redemption: "The play's struggle involves a battle to impose conformity, rather than to resist it," he writes. "*Suddenly Last Summer* is Williams's one play that resists liberation, or finds in liberation that depravity which the orthodoxy of the 1950s believed was synonymous with homosexuality." John Clum sees Sebastian Venable as a variation of *Streetcar*'s Allan Grey, a homosexual destroyed by exposure, whose homosexuality is linked not only with a "brutal, carnivorous sense of life," but with Williams's own sexual behavior, so that Clum may condemn not only Sebastian's character but Williams's as well. Further, anyone indulging in appetite to the extent Sebastian did deserves, according to Williams (as this argument continues) to be punished with equal brutality. Therefore, being devoured is a punishment that on some level Williams desired for himself.[56]

Such criticisms, made in the late 1980s and early 1990s recall comments from an earlier era. Nancy Tischler, who wrote the first book-length study of Williams, condemns the play outright. "No other play by Tennessee Williams," she wrote, "so directly calls for the adjective *sick*." And yet, the play is, she believed, a morality play that "insists on the inevitable punishment" for homosexuality: "If not preaching against abnormality, he is pointing out the natural and social consequences." It is curious the way critics writing from a stance of "out-and-proud" homosexuality can sound so much like their opposites.[57]

As he did in *Camino Real,* Williams is honestly portraying here a particular sexual appetite. Oddly, where critics blame Williams in other plays of indirectness and too firm an attachment to the closet, in *Suddenly Last Summer* it is not his closetedness that so discomfits them, it is his willingness to describe such down-and-dirty, and usually unspeakable, earthy appetites—not neat, not clean, not polite; neither "positive" nor "correct." Not every element of Sebastian's sexual life, however, is extraordinary. Clum and others accuse Sebastian and Williams of objectifying their "victims" in the basest way. Clum points to Catharine's recollection of Sebastian being "famished for blonds, he was fed up with the dark ones and was famished for blonds" and points out the similar language Williams used in a letter to Donald Windham. According to this view, *Suddenly Last Summer* is no more than a personal psychodrama of self-accusation and punishment concerning homosexual guilt. Being fed up with dark ones and famished for light ones, however, is no different from any person's description of their favorite sexual type. Williams just happens to be honest about having one.[58]

Unless one is looking very hard to find them, there is neither homophobic discourse nor indirect language in *Suddenly Last Summer.* Indeed, the play is not primarily about any image of homosexuality at all. One could as easily say that, regarding Sebastian Venable, *Suddenly Last Summer* is an unusually "positive" image of a gay man, in that its gay man has a desire to live life at its furthest edge, to engage it at its most sensual extreme, to be himself no matter what anyone else might think or how high the price of such experience may be.

Sebastian knew exactly what he was about. He believed in appetite. More than that, he believed in and accepted the death that unchecked appetite may bring. In Scene One Violet tells us that "[. . .] nothing was accidental, everything was planned and designed in Sebastian's life [. . .]" (351). He left nothing to chance, and so he may well have viewed his end as ordained. Sebastian may have been frightened by the pack of boys who turned on him, and ran from them to save himself, but still it seems clear that he and they, like Stanley and Blanche in *Streetcar,* had a date with each other from the beginning. Says Catharine in Scene Four, with an emphasis on each word, "*He!—accepted!—all!*—as—how!—things!—are! [...]* He thought it unfitting to ever take any action about anything whatsoever!—except to go on doing as something in him directed . . ." (419). Sebastian's life was one of freely chosen indulgence in appetite; his death—chaotic, sensual, earthy, violent—one of extremity of feeling. It is not the nice death his mother (who insists that Sebastian was chaste) wishes for. Sebastian sacrificed himself to his own terrible god, living and dying by his own lights.

Given the circumstances of Williams's life when he wrote the play, *Suddenly Last Summer* can also be viewed as a personal Declaration of Independence. Kubie, in the name of the principal of deprivation, called on Williams to give up writing and sex, at least temporarily. In creating a character with a sexual appetite as radical as Sebastian's, Williams suggests that, as much as a part of him might like to be saved by an ideal doctor, another part is going to defy the medical man's prescription for salvation if the patient senses that such a rescue will disrupt his true self. In this sense, *Suddenly Last Summer* is in no way a "negative image" of homosexuality. Rather, it is a full-throated defense of living one's life according to one's own lights, regardless of the consequences or of the opinions of others. While Williams was struggling with depression, his "blue demons" in Macon, Georgia, in the summer of 1942, he wrote in his journal, "I can't accept a little or nothing. I will struggle and lunge which may only tighten the bonds. I won't ever make a good captive. No, I won't make a good end of it, either. I guess what I will do is drive beyond safety—til I smash—Cleanly and completely the only hope." Such a smash-up, and a life lived to extremity, as Sebastian Venable's, is the ultimate escape, the final flight. If it is a brutal flight, if it leaves behind it a trail of exploitation, it is also one of integrity, of submitting oneself to the same exploitation at the hands of others.[59]

To say this much, however, does not permit one to overlook the fact that in his headlong devotion to his own life, Sebastian was indeed an exploiter. He used his mother, his cousin, and the hundreds of boys who crossed his path to satisfy his own desires. This habit of using others to obtain one's own end is the cannibalism that offended Williams. The play's cannibalism, as Williams was at pains many times to point out, was metaphorical, not meant to be taken literally, and it became another theme that would recur throughout the rest of his career, playing a crucial part in later plays including *Clothes for a Summer Hotel, Something Cloudy, Something Clear* and *The Notebook of Trigorin.* Sebastian is a monster not because he is homosexual, but because he is a selfish exploiter. This was precisely the guilt that Williams felt and would also project onto others, and that Donald Windham wrote about in his journal apropos *Cat on a Hot Tin Roof.* In this sense, as well, *Suddenly Last Summer* has nothing to do with guilt or loathing as far as homosexuality is concerned. This was the guilt and loathing Williams felt was over his exploitation of other people's pain.[60]

Sebastian may be the play's principal exploiter (although for that honor he contends with his mother), but he is far from the play's protagonist. That is Catharine, who is most exploited by those around her: first by Sebastian, then by her aunt Violet, then by her own mother and brother.

This appetite for others permeates every character in the play other than Catharine, and she is well aware of it. Indeed, it is Catharine who famously puts it into words: "Yes, we all use each other and that's what we think of as love, and not being able to use each other is what's— *hate. . . . "* (396). Catharine's brother and their mother implore her to lie about Sebastian's death for their sake, so that they may inherit his estate. Dr. Sugar and Violet are willing to use each other for their mutual benefit, and even the fact that Catharine tells her horrible story under the influence of a truth serum is not enough to persuade the Doctor to give up his hope of Violet's patronage. "I think we ought to at least consider the possibility that the girl's story could be true. . . ." (423) is the most he can bring himself to say.[61]

Cannibalism permeates every aspect of the play's world, including the physical. The garden in which the action occurs, according to Williams's description, is,

> *more like a tropical jungle, or forest, in the prehistoric age of giant fern-forests when living creatures had flippers turning to limbs and scales to skin. The colors of this jungle-garden are violent, especially since it is steaming with heat after rain. There are massive tree-flowers that suggest organs of a body, torn out, still glistening with undried blood; there are harsh cries and*

sibilant hissings and thrashing sounds in the garden as if it were inhabited
by beasts, serpents and birds, all of a savage nature. . . . (349)

Those like Catharine who wish to tell the truth risk having it cut out
of their brain by a surgeon's knife. The society represented by Violet,
by her greedy sister-in-law Mrs. Holly and nephew George, and by Dr.
Sugar is the same one we found in *Camino Real.* Those who oppose it
with the truth must be punished. Those with shocking appetites must
be erased and replaced with a figure whose story can, in the words of
George Holly, be told "to civilized people in a civilized up-to-date
country!" (381). In a world such as this, Catharine will be silenced and
Sebastian's image sanitized. *Suddenly Last Summer* doesn't condemn
homosexuality at all. It is a critique of a society that represses physical
appetite while encouraging greed and hypocrisy. The conflict between
the truth of people's natures and the demands of an oppressive, materialistic society based on order and control make *Suddenly Last Summer*
very much "a true story for our time and the world we live in," then
and now (382).

VII

It is easy to see the social and political aspects to *Suddenly Last Summer:*
all society is like a Venus fly trap; humans, individually and in groups,
devour their own. The principal fact, however, governing any approach
to *Suddenly Last Summer* is that, unlike any other major Williams play, it
was written while the playwright was undergoing intense psychoanalysis,
five days a week.

There are, beyond the play's social aspect, others, compelling in their
ambivalence and contradictions. Not surprisingly, these contemplate the
dynamics of the Williams family of St. Louis in the 1920s and 1930s, not
unlike *Something Unspoken. Something Unspoken,* however, is a simple
play. Beneath its plain language and unitary plot Williams plays out reflections of his parents' relationship and that between his father and himself. Beneath the baroque setting and filigreed language of *Suddenly Last*
Summer is a welter of conflicting feelings and purposes far richer than
images of homosexuals.

Violet Venable is not without appetites of her own. Both before and after Sebastian's death she used her son as much as he used her in order to give her own life a purpose and an image. Her attachment to her son had been extraordinary: If she could help it, she never let him out of her sight. When he ran off to a monastery in the Himalayas to avoid her, she froze his bank account and had him sent for. "In less than a month," she recounts triumphantly, "he got up off the filthy grass mat and threw the rice bowl away—and booked us into Shepheard's Hotel in Cairo and the Ritz in Paris—" (358). This was at the time when her husband was dying at home and desperately cabling her to return. She was quite willing to sacrifice the husband to save Sebastian; indeed, the husband (whose name doesn't seem to Violet worth mentioning) is like the spent male praying mantis that, having made Sebastian's birth possible, is now dispensable.

Violet refers to Sebastian's flight as a "crisis," but a crisis for whom? Only she could really satisfy her son, she tells Dr. Sugar. Her description of herself and Sebastian as they toured international watering-holes is almost incestuous:

> We were a famous couple. People didn't speak of Sebastian and his mother or Mrs. Venable and her son, they said "Sebastian and Violet, Violet and Sebastian are staying at the Lido" [. . .] and every appearance, every time we appeared, attention was centered on *us!*—*everyone else! Eclipsed!* (362)

After his death, Violet still tries to control Sebastian and his image before a world that doesn't know he existed. In her bitterness and rage at losing her son, Violet even manages to blame him, as well as Catharine, for his own death because he deserted her when, due to her illness, she could no longer fulfill her function in his life.

The interior of Mrs. Venable's house is blended, the set description says, with Sebastian's garden, suggesting not just society as a jungle but households as jungles, as well. Not only do we, as a society, feed on each other, but mothers, fathers, sisters, and brothers are equally susceptible to voracious appetites that can best be satisfied by dining on their nearest and dearest.

A psychologically minded critic will prick up his or her ears at the mention of Violet's house and Sebastian's garden blending together. At a certain point they are indistinguishable. A psychoanalytically minded critic, or, for that matter, a Freudian psychoanalyst in 1958, might say this is evidence of a child who never established an identity separate from his mother and unconsciously fears being devoured by her. In Violet's case, the fear is not unfounded. Violet is certainly the devourer of Sebastian. "I would say 'You *will!*' and he *would* [. . .]" Violet tells Dr. Sugar proudly (408). She was not only his mother, she was—according to her—his muse, as well. Only she, Violet believes, could make certain that Sebastian would write his annual *Poem of Summer*.[62]

If, in Williams's mind, anyway, Edwina was Rose's executioner, she threatened to be his extinguisher. The story of what happened to Sebastian in Cabeza de Lobo started, Catharine tells Dr. Sugar, "the day he was born in this house." And yet, perhaps, Sebastian feared an emotional break with his mother. After Violet's stroke rendered her useless to him, the poet found he could not write his poem. Something in him had broken, Catharine says, "that string of pearls that old mothers hold their sons by like a—sort of a—sort of—*umbilical* cord, *long— after . . .*" (408–9).

Violet is not the only mother to act as an exploiter, however. Catharine's mother has not been to visit Catharine in the mental hospital where she is currently confined. In fact, she assures her daughter, she has gone to great lengths to ensure that nobody knows what has happened to her or where she is. It's hard to see how the fact that no one knows of her unjust incarceration in a mental institution could offer Catharine much comfort. How perfectly, however, this suits the purposes of Mrs. Holly and her son George, who want nothing more than to appease Violet and so get Sebastian's will, from which they will profit substantially, out of probate.

Customarily, Catharine has been equated with Rose Williams, and the reasons are obvious. In the context of Williams's daily visits to Kubie, however, the character might also be seen as the playwright's view of himself, hoping that the intervention of an idealized doctor will save him from the predations of a fearsome, castrating mother, the violent images of his imagination, and his exploitative, and not incidentally lucrative, ways. If, as Violet says, Sebastian's life was his work and his work was life, then his life consisted of devouring others for his own purposes, and this is how Williams had long felt about himself. Thus, to be saved by Dr. Sugar/Kubie meant the end of pain and the beginning of peace. For Williams, the stakes were as high as they are for Violet Venable and Catharine.

Yet, Williams could not have been anything but ambivalent toward being saved by Dr. Sugar/Kubie if salvation meant ending exploitation, for this would mean giving up writing. So perhaps the play could not end on anything but an ambiguous note. To give up writing was the one thing Williams was never able to do; he could temporarily give up alcohol and pills, but writing was the one addiction he never conquered. However, if Williams was to remain a writer—despite Kubie's suggestions to do otherwise—what choice had he but to continue to exploit the sufferings of all his family members? His career and daily existence depended on using other people—so often Rose, but others, too, including his parents and his brother Dakin (who appears as the lunkish lawyer Gooper, Brick's brother, in *Cat*). What way out was there? For in writing this play about the horror of using other people and calling it love, he was forced to use them again. Thus, Edwina, in the guise of Mrs. Venable, not only calls for the mental and spiritual destruction of Catharine (her daughter and her son), but at the same time, Williams makes Violet the *champion* of her writing son, the one person on earth dedicated to preserving his memory, and the only one who could induce him to write his annual poem. This was a paradox Williams was unable, and perhaps unwilling, to resolve in therapy with Kubie. As for sex and the saving feeling that physical intimacy gave him, giving that up, that, too, would be impossible. So, like Sebastian Venable, Williams would go on doing as something in him directed, even if it meant sacrificing himself to his terrible God.[63]

Kubie saw *Garden District* and came away deeply impressed by the longer play on the bill. Shortly after, he wrote Williams a letter. "My Dear Mr. Williams," he begins in the letter dated January 13, 1958:

> I am still very deeply under the influence of the remarkable experience of seeing your plays on Thursday night. All three of us felt the spell. They are

remarkable not for the beauty of your writing alone, but also for their many subtle implications and partial but penetrating insights into some of the fundamental problems which convulse the human spirit.[64]

Regarding *Something Unspoken*, Kubie wished that Williams had concentrated entirely on the relationship between Cornelia and Grace. Spending so much time on Cornelia's attempt to be acclaimed Regent detracted, he felt, from the piece and dragged it down into the realm of burlesque.

After praising the principal women of *Suddenly Last Summer*, Kubie mentions his particular pleasure at Robert Lansing's portrayal of Dr. Sugar:

> I wish you would tell him someday that of the many portrayals of the role of the psychiatrist that I have seen on stage and film, his rang truest. It had a quality of thoughtful, unpretentious, [sic] competence, of responsibility and humanity. He had strength without having to make a show of it; and he did not have bedside manner oozing out of every pore. It was good.

He closes by hoping Williams is having a good holiday and wishes that he finds "the inner peace that you are seeking." It may be that Kubie was doing a little projecting: nowhere in the play does Williams identify Cukrowicz, a neurosurgeon, as a psychiatrist as well. Nowhere in his letter does Kubie make any sort of judgment of Sebastian's homosexuality or connect it to the manner of his death." In a postscript he adds, "Please extend my greetings to Frank."

VIII

Williams chose to produce *Garden District* Off-Broadway because, he told *The New York Times*, he didn't want to attack Broadway audiences. A year later, he wrote in the *Times* that he'd expected to be "critically tarred and feathered and ridden on a fence rail out of the New York theatre," for perpetrating *Suddenly Last Summer*. He needn't have worried. The reviews were almost unanimously positive. The unsigned critic from *Cue* magazine found the topic of homosexuality rather old-hat and greeted the play with a yawn, then with a dismissal: "Everybody, it seems, has written of homosexuality during the past decade. It's a fashionable and best-selling subject. But perhaps the shock value is wearing thin." Most critics praised the play more highly than any Williams work since *Streetcar*, and did so without mentioning Sebastian's sexuality, the manner of his death, or even the play's condemnation of society's exploiters. The leading Boston critic, Elliot Norton, couldn't bring himself to say that Sebastian was a homosexual in his review in the *Boston Daily Record*. The young man is simply "eccentric" and "strange." In one of the most positive reviews Williams would ever receive from a daily New York paper, Richard Watts in the *Post* also refrained from mentioning Sebastian's homosexuality—or any other specific detail of the play.[65]

While most of the other New York critics found a way to say "cannibalism" in their reviews, none except the critics for *Cue* and *The Saturday Review* could manage "homosexual." For all of its familiarity on Broadway through the 1950s, and the less formal, perhaps more permissive Off-Broadway setting Williams had chosen for his new play, homosexuality was still a topic about which critics were reluctant to speak. Brooks Atkinson, who, in addition to his review wrote a longer appreciation a few days later, and who was clearly deeply affected by the play, referred to Sebastian as a sybarite and an Epicurean; for George Oppenheimer in *Newsday*, he was a parsimonious poet; in *Time* he was self-luxuriating one and in *Theatre Arts*, where one might have expected the editors to recognize the readers who comprised a significant part of their more specialized audience, Sebastian was simply dead. *The Saturday Review* lumped Sebastian's homosexuality and his death together: Williams, Henry Hewes wrote, once again had surrounded his characters in a "jungle of homosexuality, alcoholism, masochism, sadism, and now cannibalism. . . ." Thanks to the critics' refusal to acknowledge the characteristics of the figure around whom *Suddenly Last Summer* revolves, it would have been entirely possible to purchase a ticket to *Garden District* on the strength of its reviews and not know anything specific about Sebastian Venable other than that he was dead.[66]

A few years after *Suddenly Last Summer*, Edwina would become an author herself, using her son and daughter as material for her own gain. What could Williams do but laugh, particularly while reading Edwina's quick summary of his career: "Murder, cannibalism, castration, madness, incest, rape, adultery, nymphomania, homosexuality. There exists no savage act about which my son has not written."[67]

FIVE

Almost Willfully Out of Contact With the World

Many in the gay theatre-going audiences had always known of Williams's homosexuality; he had never hidden his private life. Donald Vining reflected the degree to which gay audiences had embraced Williams as an icon when he wrote in his diary about the large audience of queer men who cheered at the curtain calls for *The Rose Tattoo* in 1951. In 1968, when Mart Crowley included references to *Suddenly Last Summer* in *The Boys in the Band*, he felt no need to explain them. This embrace of Williams by a New York gay audience would not last, however, very much beyond Crowley's play. Changes in the social and political climate, as well as in coverage of gay playwrights in the New York press, and, most importantly, changes in the attitudes of gay men and women themselves, would turn Williams into a figure of ridicule, when not ignored altogether, for many in the emerging gay community of the early 1970s.

For most of his life, Williams never identified himself as a gay playwright, if only because such an identity did not exist for homosexual men and women of his generation. He never hid his homosexuality from his professional colleagues, but, as was usually the case for gay men born in the first years of the twentieth century, Williams never identified himself primarily as gay. Neither did he write gay plays, that is, plays that sprang from specifically gay experiences and took them as their central concern. Nor did he feel the need to create positive images of gay men anymore than he created such politically based images of any other kind of character.

Many lesbians and gay men who came of age in the 1960s experienced their sexuality differently than people of Williams's generation had and this new era of gay theatre artists was reflected by the difference in their work. Many more were able to embrace their homosexuality proudly and publicly. They often identified themselves as gay men or women, or as gay writers; and those involved in the gay theatre that emerged in the 1960s and bloomed in the 1970s and early 1980s were likely to view theatre as a useful tool for community-building and for disseminating sexual and politically positive images of themselves. It was inevitable that they and Williams would misunderstand each other.

Small Craft Warnings, produced in 1972, was therefore taken by some as Williams's first utterance of what it meant to be a gay man in the post-Stonewall era. To those who saw it that way, it must have appeared appallingly out of date. The play, however, attempted to make no such statement. It was, rather, a report on the playwright's psychic and spiritual condition over a decade beginning about 1962. What Williams offered as a personal meditation on his own life would be taken by many, including gay critics writing years later, as a homophobic portrait of gay men. What had happened in American culture—specifically theatre—and society to create such a chasm between Williams's intent and a younger generation's perceptions?

I

For Tennessee Williams, there had always been only Broadway. Whatever he had to say he put in such a way so as to attract not only a general audience large enough to sustain a commercial run, but, more fundamentally, to entice producers who would be willing to risk their own and other people's money on the stories Williams had to tell. Even within these relatively confining commercial conditions, however, Williams, through the creative tension spurred by his dual desires to reveal and to conceal, was able to write what he considered to be the truth without resorting to distortions such as disguising gay men as heterosexuals. By the early 1960s, another theatre was developing in New York: a gay theatre, in which writers achieved a freedom of expression far beyond that of Broadway, where the need to conceal any kind of physical desire was negated by a burgeoning youth culture that was prepared to embrace any sexual revelation as an aspect of its own identity.

If the New York gay theatre could be said to have a definitive beginning, it would be in December 1958, when an ex-dancer named Joe Cino opened a coffeehouse at 31 Cornelia Street in Greenwich Village called The Caffe Cino. At the beginning, Cino had no intention of producing plays. After presenting a few poetry readings, however, so many writers and actors approached him about putting on plays that soon he was mounting small productions. Through Cino's death in 1967, Caffe Cino produced a new play every week or two. While the Cino soon became known as the birthplace for new American plays by Sam Shepard, Lanford Wilson, and others, its initial productions were also important contributions to the embryonic gay theatre movement. These included André Gide's version of *Philoctetes,* Oscar Wilde's *Salome* and *The Importance of Being Earnest,* Jean Genet's *Deathwatch,* and several one-acts by Tennessee Williams.[1]

What Cino and others were establishing was a movement bent on creating an alternative to the commercially driven theatre of Broadway and even to the less buttoned-up, still-emerging Off-Broadway, of which Williams had been a pioneer with *Garden District.* Although the movement was not specifically gay, many gay artists were involved and they began expressing aspects of their gayness through their work in this collection of theatre spaces, mostly in the East and West Villages, which became known as Off Off-Broadway. In 1960, in a basement on East 12th Street, Ellen Stewart established Cafe La Mama as a private club, where members paid weekly dues of one dollar that entitled them to see the current production as many times as they pleased. One of her first productions was an adaptation of Williams's story of a male prostitute on death row, "One Arm." The following year, Judson Memorial Church, on Washington Square, initiated several arts programs that were conceived as being integral parts of the church's work. The Judson Poets' Theatre, as the dramatic arm was called, was presided over by the church's assistant minister, Al Carmines. The members of the loosely organized theatre company collaborated closely with church members, who voted never to censor a play's content or language. Productions were produced in the choir loft and in the nave, sometimes in collaboration with the church's other landmark arts organization, the Judson Dance Theatre. In 1963, with profits from of *Who's Afraid of Virginia Woolf?*, Edward Albee and the play's producers, Clinton Wilder and Richard Barr, established the Playwrights Unit. It presented about one hundred plays in eight years, many by the playwrights who began at the Cino, La Mama, Judson Poets' Theatre, and other Off Off-Broadway venues.[2]

One certainly did not have to be gay to be produced at an Off Off-Broadway theatre, nor did the subject matter have to be. Many of Joe Cino's patrons were gay (as was he), however, and the theatre that Cino produced often reflected their sensibilities, concerns, and styles: In addition to Lanford Wilson, gay playwrights who saw their plays premier at the Cino included Tom Eyen, Robert Heide, William Hoffman, H. M. Koutoukas, Robert Patrick, Jeff Weiss, Doric Wilson, and David Starkweather. (In later years, high school and community theatre directors might have been dismayed had they known that one of their staple attractions, the musical movie spoof *Dames at Sea,* a work very much in the emerging gay spirit, began at the Cino in 1966.) While Stewart, at La Mama, like Cino, hardly had a formal policy of producing plays by gay writers, many of the playwrights whose work she put on at her cafe were gay, including Eyen, Jean-Claude van Itallie, Megan Terry, Lanford and Doric Wilson, and others. Neither was Judson Poets' Theatre specifically gay, but many of its participants were, and a number of its productions, including Carmines's musical *Faggots,* certainly derived from a gay sensibility. Indeed, when the Poets' Theatre's resident director, Lawrence Kornfeld, was accused of producing a "campy" show (Ron Tavel's *Gorilla*

Queen in 1967) he responded, "Campy, hell! It's downright homosexual!" Hovering over the Judson enterprise was the spirit of Gertrude Stein, several of whose plays Carmines produced, including musicalized versions of *What Happened,* which became the theatre's first notable success, and *A Circular Play.*[3]

Some of the plays produced at these theatres had gay characters. More of them were gay in style and tone. Like homosexuals themselves, "gay style" may be hard to define and easy to know when you see it. William Hoffman, in his important collection of gay plays, speaks of the difference between "gay plays" and "gay theatre." A "gay play" may have gay characters at its center and concern itself with situations and themes of specific concern to gay people and springing from their experience. "Gay theatre," on the other hand, may take any topic or character as its central concern, but present it in such a way as to acknowledge "that there are homosexuals on both sides of the footlights." Gay theatre, Hoffman explains, regards its audience the way two homosexuals might regard each other at a party, or on the street: with knowing nods and winks, a flirtatiousness that might be either outrageously straightforward or composed of double entendres and hidden meanings. A comic, totally unpretentious precursor to postmodernism, the gay theatre that emerged in the mid-1960s bumped comedy against tragedy, quoted works from literature, comic books and film, could be heavily ironic, was aware of itself as theatre and made its theatricality part of its subject-matter.[4]

This gay theatre reached its first apogee with the productions of John Vaccaro's Play-House of the Ridiculous, which produced the early plays of Tavel and Charles Ludlam, beginning in 1966. When Vaccaro and Ludlum quarreled midway through rehearsals of Ludlam's *Conquest of the Universe,* Ludlam established his own theatre, the Ridiculous Theatrical Company. Retitling the play *When Queens Collide,* Ludlam staged it himself, and audiences could choose between his production and Vaccaro's, at the latter's newly renamed Theatre of the Ridiculous.[5]

These theatres and the others that established Off Off-Broadway as both a geographical entity and an elastic but generally shared aesthetic movement (with an emphasis on the playwrights' words and usually the simplest of production values), had nothing to do with the values of the theatre uptown. They could be found in lofts, storefronts, bars; the usually cramped quarters of these spaces, where audiences were often no more than a few feet away from performers, lent productions an intimacy and informality that was totally foreign to the Broadway theatergoer's experience. Michael Smith, a young critic for *The Village Voice* when Off Off-Broadway was emerging, and was one of its first critical champions, defined the force behind these theatres: "To work freely toward your own vision, to set your own standards, to define your own goals—the by-words of Off Off-Broadway are inconceivable on Broadway."[6]

What gave birth to these theatres was neither the commercial instinct nor, in most cases, a desire for fame. The theatres that became Off Off-Broadway developed as the voices of a fairly homogenous community ready to make its voice heard in the public spaces that happened to be available to it. The idea of performance in a coffeehouse was not new; thanks to the Beat poets and musicians, such places in Greenwich Village had been homes to poetry readings and musical performance for years. The community of young writers and actors who emerged in the late 1950s and early 1960s found places to produce their work and an inexpensive, highly theatrical aesthetic in which to stage it for an audience very much like themselves, who lived in the neighborhood and shared their political, aesthetic, and social concerns. Not only could productions be mounted cheaply and charge low admission fees (the double-bill that opened the Judson Poets' Theatre in 1961 cost $37.50), but artists were able to subsist on low-paying jobs, pay their rent, and still have the time and energy to create theatre.[7]

Off Off-Broadway was that rarest of theatres: an authentic expression of a relatively homogenous community that grew up organically, thanks to a specific set of economic and social circumstances. Some members of this creative community, such as The Living Theatre and the Open Theatre, were interested in developing radically different forms of theatre and changing the way audiences perceived it. For others, it was a theatre whose creators wanted to show their lives frankly and naturally and saw no reason not to. Perhaps this was the theatre in which Tennessee Williams was meant to work all along.[8]

Broadway theatres produce a specific set of expectations in audiences, reinforced by the theatres' formal layout and appearance, including plush red seats (a patron could sit only in the seat his or her ticket indicated and every seat was reserved), formal ushers, and the high proscenium picture-frame stage dominated by a majestic curtain that was separated from the audience by the gulf of an orchestra pit. The decor tends to be classical, symmetrical, Beaux-Art. Everything about a performance in a Broadway theatre suggests a formality, a conservatism, a distance and, until the late 1960s, this atmosphere was reinforced by the middle- and upper-class audience's habit of dressing for the theatre in business or even formal attire. None of Williams's plays after *The Night of the Iguana* was suited for such an atmosphere or audience.

As a home for his work, Williams was at best ambivalent, and at worst dismissive, of Off-Broadway (let alone Off Off-Broadway). In 1961, seven weeks prior to the opening of *The Night of the Iguana,* he told *The New York Times* that he was tired of Broadway and no longer up to the strain that an uptown opening invariably produced in him. "I don't think these bouts are good for me," he told the writer Lewis Funke. "Nor do I find working on Broadway any fun." He acknowledged the successes that Off-Broadway had brought him (six years prior to *Suddenly Last Summer,* José Quintero and the new Circle-in-the-Square had revived *Summer and Smoke* and made a star of the young Geraldine Page and burnished the play's reputation) and he even predicted that a new play, *The Milk Train Doesn't Stop Here Anymore,* might be a good candidate for the smaller, more intimate circumstances of an Off-Broadway theatre. When *Milk Train* opened, however, first in 1963 and in a revised form a year later, it was on Broadway, and both times the productions closed with disastrous quickness. *Milk Train* was an outré work in which Williams replaced Western theatre conventions with those of Japanese Noh. Rejecting his well-known (and to him, increasingly well-worn) theme of survival in favor of a new belief in the value of resignation, the play might have found an audience interested in adventurous theatre in a downtown performing space.[9]

In 1970, Williams spoke derisively of Audrey Wood's supposed desire to produce *The Two-Character Play* Off-Broadway. "I think she wants *all* my work produced Off-Broadway," he said. Had Wood actually advocated producing the play Off-Broadway and prevailed, it might not have been dismissed as curtly as it was by both critics and audiences. Yet by reputation and habit Williams remained a Broadway playwright as long as there were producers willing to risk increasingly large sums on his plays. He was used to working with star actors and directors who, he believed, would not work for Off-Broadway salaries—let alone for free, which was usually the case in Off Off-Broadway. As much as Williams applauded Quintero's production of *Summer and Smoke*, he failed to appreciate the role that the informal, intimate theatre, in which the audience sat close to the action and surrounded it on three sides, played in the production's success. It may be that, like Brick Pollitt, Williams was unable to give up the visible signs of privilege and fame that Broadway bestowed on its successes.[10]

After *Suddenly Last Summer*, Williams would not present new work in New York away from Broadway until the early 1980s. Then, he was forced to find venues elsewhere because producers were no longer willing to risk large sums on a playwright whose new Broadway work had not made a profit in 20 years. In the early 1970s, his plays would be produced at the Off-Off Broadway Truck and Warehouse Theatre (*Small Craft Warnings*) and the Hudson Guild (*A Lovely Sunday for Creve Coeur*) and they were respectfully, if unspectacularly, received. He insisted on Broadway productions for *The Red Devil Battery Sign*, *Vieux Carré*, and *Clothes for a Summer Hotel* and they failed to find an audience.

Meanwhile, the important work that would shape American theatre for the rest of the century was occurring in coffeehouses, cafes, churches, and bars. The notion of theatre done cheaply, on the fly, for a local audience by people who were members of that community, had spread to Chicago where a healthy network of theatres sprang up, including Second City, whose alumni, in the coming decade, would work in television and film as well as theatre. The urge of a community to express its voice through theatre without regard for uptown values or opinions, which in New York produced Shepard, Wilson, Rosalyn Drexler, Terry, and so many others, would eventually give birth to the plays of David Mamet in Chicago. The gay aesthetic, an increasingly important and visible aspect of this movement, would flower and become intentional and more overtly political. In the next decade, when Williams returned to writing gay characters, increasingly putting them in the center of his plays, what he was doing would strike many gay theatre-makers and activists not only as old-hat but also not nearly gay enough. For them, Tennessee Williams would be offering much too little far too late.

II

While America and American theatre witnessed seismic cultural and political upheavals between 1958 and 1972, where was Tennessee Williams? To understand *Small Craft Warnings* and the plays that came after, we have to trace the journey he took to the bottom of his soul in those years. Williams viewed whatever was going on beyond his own contracting world through an increasingly distorted and distancing lens. His need for both alcohol and pills had only increased after he'd unleashed the taut psychic violence of *Suddenly Last Summer*. By 1962, he was telling *Theatre Arts* magazine that only swimming, liquor, and the tranquilizer Miltown released him from his habitual restlessness. Writing was not on the list, although he stuck faithfully to his habit of working every morning.[11]

At the end of 1961, Williams saw what would be his last commercial success open on Broadway. *The Night of the Iguana* is one of Williams's few plays that ends on a note of grace and hope, a fact all the more remarkable since it was written under circumstances that were, at least up until this time, more harrowing than usual. His paranoia had been growing unabated, and he began cutting off old friends, beginning with the loyal Cheryl Crawford, who had produced *The Rose Tattoo, Camino Real, Sweet Bird of Youth,* and, in 1960, *Period of Adjustment* and who'd hoped to produce *Iguana* as well. After *Period of Adjustment* closed, after 132 performances, Williams simply stopped speaking to her. His relationship with Audrey Wood, more faithful to him, perhaps, than anyone except Frank Merlo, also began to cool. In an article written for *Esquire* about the author-agent relationship, Williams admitted feeling resentful, smothered, and humiliated by a person making too many business decisions for him. He did not mention that had he made any of those decisions himself, he would not have had to rely on Wood so completely.[12]

When rehearsals began for *Iguana*, he would deliver daily rewrites as usual, but they were not bringing the play into sharper focus. The endless amount of new lines only contributed tension to a troubled rehearsal period in which the star, Bette Davis, clashed frequently with director Frank Corsaro. Eventually, she demanded that Corsaro be fired and take his Actors Studio sensibilities with him.

While the play was trying out in Detroit, an over-drugged Williams was bitten on the ankle by his bulldog Satan. The wound became infected, Williams had to be hospitalized and he irrationally accused Merlo of setting the dog on him. Merlo, he said, wanted him dead. This relationship, too, which had been the major stabilizing force in his life since 1948, and which was already weakened, soon would be damaged beyond repair. Williams managed to right himself long enough to return to work on *Iguana* and see it through its successful Broadway opening, but between him and Merlo, things would never be the same. They never formally split up, but began spending less and less time together. Instead, Williams sought the companionship of a number of men, none of whom proved to be as devoted, caring, or patient as Merlo.

Williams knew that his increasingly irrational behavior was leading to an irreparable breach even as he seemed helpless to change. After a quarrel at the end of 1960, Williams had written Merlo a letter that was at once conciliatory and manipulative. "Dear Horse, or St. Francis," it began.

> I guess you win, like Mizzou in the Orange Bowl Game. Thirteen years, the longest war on record, but that's not a nice way to put it. . . .
>
> I hope to behave as my father did when he lost but I hope that unlike him, I won't be locked out of the house. I have no Knoxville to go back to, and no widow from Toledo.
>
> If it should not turn out to be an honorable capitulation, I suppose I could still employ a traveling companion who would take me away to Europe but then the victory, yours, would lose its glory and even its just reward, and I do mean just because I think to have passed thirteen years with me, the gloomy hun of all time, must merit a crown in heaven.
>
> Blanche was a bit of a hun, too, but I think she was quite sincere when she said, "Thank you for being so kind, I need kindness now."
>
> Kindness in one makes kindness in the other.
>
> Love,
> T[13]

Some of the changes in their relationship were the result of Williams's evolving financial circumstances. As he had grown wealthier, he no longer needed Merlo to be major domo, cook, social secretary or to fulfill any of the other roles Merlo had played in keeping Williams's daily life in order. Williams began hiring people to do this work, and it didn't occur to him that Merlo had performed these services because they fulfilled him, and because he loved Williams. As he was gradually deprived of his central role in Williams's life, Merlo began to feel useless.[14]

Williams had met a young poet, Frederick Nicklaus, and invited him to move into the house on Duncan Street in Key West. Although Merlo's and Williams's relationship was not what it had been and both had been sleeping with other men, Merlo understandably resented Nicklaus's presence, and his anger was exacerbated by the frustration caused by his enforced inactivity. A four-pack-a-day smoker, he had been feeling increasingly ill and weak for several months, and his condition was not improved by the new living arrangements in Key West.

When Merlo's condition worsened, Williams rushed to Key West from London, where he and Nicklaus were vacationing after the Spoleto premiere of *The Milk Train Doesn't Stop Here Anymore*. When Williams went to New York in October to work on *Milk Train*'s Broadway opening, he kept tabs on Merlo's condition, and the cast sensed that his mind was not on his work.

Milk Train opened in January 1963, to poor, and in some cases, cruel reviews. Richard Gilman, an influential young critic, titled his review in "Mistuh Williams, He Dead," and declared, "Why, rather than be banal and hysterical and absurd, doesn't he keep quiet?" That same month, Merlo came to New York to consult a physician about his hacking cough and weight loss. The doctor told him his condition was inoperable lung cancer.[15]

In August, Merlo was admitted to Memorial Hospital in New York. Williams wrote Nicklaus in Key West that Merlo was experiencing both severe pain and mental distress. The situation, he said, was appalling. Williams visited every day that month, and it was clear to the family of Merlo's hospital roommate that Frank's worsening condition was deeply upsetting to Williams. In September, perhaps as much for a break as anything else, Williams went to Abingdon, Virginia, for a production of a revised *Milktrain,* which was scheduled to reopen on Broadway the following January. He returned to New York on September 20. Late the next evening, after visiting hours had ended and Williams had gone for a drink with friends, Merlo died. He was 41.[16]

On the 23rd, Williams wrote Maria St. Just, "He died proudly and stoically; they tell me it happened in a matter of minutes, before the floor-doctor had time to reach him, but there was a nurse in his room who had just given him his nightly medication. They say he just gasped and lay back on his pillow and was gone. I am just beginning, now, to feel the desolation of losing my dear little Horse."[17]

Williams was suddenly without the partner who for 13 years had made his life possible in ways from the most quotidian to the profound. Williams did not know where to turn. He wrote Maria St. Just that he was uncertain where to go: to London, for yet another production of *Milktrain?* To Mexico, where John Huston was filming *The Night of the Iguana?* One place he could not return to, he knew, was Key West, where he and Merlo had been as happy as Williams could be.

Now, despite the presence of any number of companions and paid assistants, Williams's life spiraled downward into a depression that would last seven years. In order to write, Williams began injecting himself with speed every morning, obtained from Max Jacobson, the infamous "Dr. Feelgood" who specialized in giving his celebrity clients shots of amphetamines. In the afternoon Williams took a variety of barbiturates, including Nembutal, Luminal, Seconal, and phenobarbital, as well as the non-barbiturate sedative Doriden. Against Jacobson's advice, he also continued drinking. Since 1962, he told *Time* magazine, he drank half a fifth of liquor a day, as well as half a Dexamyl to "pep up," one and a half Seconals to "smooth things over," and "two Miltowns with scotch to go to sleep. And when I suffer from acute insomnia, which is also often, I take up to four sleeping pills." He was deep into what he came to call his "Stoned Age."[18]

He could thank the iron constitutions of the Williamses and Dakins that he survived at all. The result for his writing was, predictably, uneven and, for his career, disastrous. The second Broadway production of *Milktrain* opened in January 1964, and was received with no more critical enthusiasm or understanding than the first. *Slapstick Tragedy* in 1966, *Outcry,* produced in London in 1967, and *Kingdom of Earth (The Seven Descents of Myrtle)* in 1968, each failed, and critics saw in them the steep, shocking demise of America's foremost playwright. Most calamitous for Williams, personally as well as professionally, was the critical response to *In the Bar of a Tokyo Hotel* in 1969. Few understood the play, which, like *Milk Train,* is based thematically and structurally in Noh theatre. Some critics, notably Clive Barnes of the *Times* and Williams's old colleague Harold Clurman in *The Nation,* were sympathetic, while others took the opportunity of announcing Williams's artistic demise. In an *ad hominem* attack cast as a review entitled "White Dwarf's Tragic Fadeout," in *Life* magazine, Stefan Kanfer wrote, "Tennessee Williams appears to be a White Dwarf. We are still receiving his messages, but it is now obvious that they come from a cinder. . . . Other playwrights have progressed: Williams has suffered an infantile regression from which there seems no exit." A week later, *Life* quoted Kanfer's review in a full-page ad in *The New York Times.* The ad comments, "That's the kind of play it is, and that's the kind of play it gets in this week's *Life.* From a theatre review that predicts the demise of one of America's major playwrights to a newsbreaking story that unseats a Supreme Court judge, we call a bad play as we see it."[19]

Williams was now convinced that people were plotting to kill him. Finally, in September 1969, his brother Dakin admitted Williams to Barnes Hospital in St. Louis. There, doctors placed him in the psychiatric division, where he stayed for two months. He was taken off his drugs cold turkey, which resulted in three grand mal seizures and two heart attacks over the course of three days. Nonetheless, the treatment saved his life. He was back in Key West by December, and while he continued to drink and take drugs (mostly sleeping pills) for the rest of his life, he had reached and passed his physical nadir. He had weathered a violent storm, largely of his own creation, and emerged more stable. He cut down on his drinking, limiting himself to one martini a day and perhaps some wine with dinner, but never entirely cut out the pills.

The plays written during Williams's "Stoned Age" dealt increasingly with death. Death had never been far from his mind, whatever his previous plays' immediate concerns, but in earlier works, such as *Orpheus Descending* and *Sweet Bird of Youth,* death was seen through the bright, lurid colors of melodrama. A martyr's death, like Lady's, Val's, or Chance's, was an intense, even sensual, experience. Survival was to be preferred, but in these plays, death could award one a spiritual victory. In *Milk Train,* it is clear that death had replaced survival in Williams's imagination as the preferable solution to guilt and suffering. In *Milk Train,* death is a release: unavoidable, natural and welcome, and the only way off the wheel of earthly suffering. In the Stoned Age plays after *Milk Train,* however, death is black, silent, devouring. It was the ultimate negation, a destiny from which there was no escape. In these plays, Death is the ultimate suffocation and a metaphor for Williams's darkest personal fear: claustrophobia and the inability to flee.

Few plays exemplify that dark vacuum more than a one-act Williams wrote during 1969 and 1970 called *Now and at the Hour of Our Death,* at about the same time he was working on *Tokyo Hotel.* It takes place in a city like New York, in a restaurant near a huge department store called Guffel's. Outside, the entire planet is sinking into chaos and death. Although no one can be seen in the street through the restaurant's main window, we're told of hordes of people everywhere. In the store, the after-Christmas sale prices create panic among the crowd of shoppers, suggesting either chronic shortages or rampant greed. "Howling confrontations, furious collisions. Purchases disarrayed if not completely destroyed," reports Bea, a shopper who escaped the mob by running down a fire escape and then jumping two stories into the crowded street. Taxicabs are impossible to hail, and the cross-town traffic has turned "The Mayor's pubic hair white!" The restaurant itself is overflowing with customers, although at first we see only two, Bea and her friend Madge, who are meeting for lunch. They speak of polluted water and oil slicks despoiling their favorite vacation haunts and "plantan" in the ocean that is "dying or even already dead," threatening the earth's supply of oxygen. A war rages overseas, in which Madge's son has been killed.[20]

The two women are as competitive as the street hordes. Williams's stage directions note that both are regarded in their circle as "*intellectuals*" and so a *"guarded rivalry"* exists between them. Both are grotesques: Madge carries an enormous purse, *"the size of a piece of luggage"* (1), and constantly cleans her glasses with a hissing spray bottle. Bea enters with a giant plush rabbit toy, a birthday gift for the mongoloid daughter of a friend. Neither can help feeling superior and making snide remarks about the other's appearance, marital situation, or place in a world that is deteriorating at an alarming rate.

Their waitress is a victim of senseless violence: She was attacked by a stranger that morning in the subway. Now she sports a black eye and although she is traumatized almost to the point of catatonia, Bea and Madge mock her. "Watch out for those bathroom doors, dear," Madge jokes, assuming she's been beaten by her boyfriend or husband (6).

Visible through the window, a Hunched Man in Black appears carrying a placard. One side reads, "Clowns Precede," the other, "The Appearance of Tragedians" (8). In his stage directions, Williams has already referred to the women as "the clowns," while the two men whose entrance the Hunched Man heralds are "the major figures (tragedians) [. . .]" (2).

The tragedians enter to an offstage trumpet call. They are Dave and Jack, gay lovers. Dressed in leather motorcycle jackets with the inscription, "The Mystic Rose" on the shoulder, they are "strikingly handsome and very young [. . .]" (21). We instantly notice a difference between them and the two women: Dave and Jack are solicitous of the waitress's condition, sympathetic to her plight. They recognize that her pain is in some sense their own. While the two clowns talk past each other and try to trump the other's cattiness, Dave and Jack listen to each other and explain, sometimes with difficulty, their feelings. In a world where the grotesque is the normal, they are truly grotesque for being natural and human.

They are also hustlers—another black mark against Williams, as far as many of his gay critics would be concerned. One hustles on the street, another out of an establishment called "Mother Freddie's." Neither likes the life, and they have a curious reason for living it: "We can't explain to no one a job would separate us," Dave says (27).

After their small meal, Dave and Jack return to the street. A few minutes later, the sound of a crash is heard. Madge looks out the window and reports, "The motor-cycle came a cropper. The brains of one of the boys are scattered on the street. The other is leaning in shock against the wall of Guffel's" (34). Quickly, employees of the store scatter sawdust over the brains and then shovel them into a barrel—reminiscent of the Baron's demise at the hands of the Streetcleaners in *Camino Real*. Dave, the survivor, staggers in and is immediately surrounded by female shoppers who, rather than offer help, pepper him questions and comments about the accident, before swiftly decamping through the restaurant's revolving door.

While by no means perfect, *Now and at the Hour of Our Death* is riveting in its circus atmosphere and outré characters. It is also haunting; it stays in the mind long after reading. Williams increases the feeling of death and decay with the strangulated, jagged dialogue he was experimenting with at this time in *In the Bar of a Tokyo Hotel*. While some critics regard that language as an attempt to further hide his homosexuality (which Williams never did in the first place), it feels more like the language first employed by the German Expressionists at the beginning of the twentieth century. There, the purpose of dialogue delivered in staccato bursts was to emphasize the difficulty of verbalizing deep feelings.[21]

Most interesting are Dave and Jack. The meaning of the insignia on their jackets—The Mystic Rose—is as much a mystery to them as to others. Asked by a patron in the restaurant's men's room what it means, Dave is unable to answer. He says to Jack, "[. . .] we need some publicity about the club to explain our thing to the public [. . . .]" Jack replies, "Aren't we still trying to find out for ourselves exactly what our thing is?" (25) A few minutes later, Jack hesitates when Dave suggests they leave. "Man," Dave says, "they're out there waiting for us."

> Jack: I don't think they'd wait. Look. They accept us, they tolerate us, but that's as far as it goes.
> Dave: We said wait a little for us, they said they'd wait, they're waiting.
> Jack: Don't you know that they know what we are?
> Dave: What are we? That they'd put down?
> Jack: A pair of human commodities, like I said.
> Dave: Man, in this world guilt feelings are—unnecessary—excess! (30)

Who are "they"? We're never told. What is it that they tolerate? The fact that Jack and Dave are gay? Written around the time of the Stonewall

Riots, the characters may be reflecting on the tentative way the straight world regards them. However they, themselves, are also trying to figure out what their "thing" is. We're not certain what that "thing" is, either: What it means to be young, gay, and out? What it means to be a commodity? Men had been commodities in Williams plays at least as early as *Talk to Me Like the Rain* written about 1950 (and produced on PBS in 1970); it will recur in *Something Cloudy, Something Clear* and in poems such as "The Blond Mediterraneans" in 1980. In Williams's work of the late 1960s, the world is a strange, frightening field where identity is uncertain, the present horrific, and the future unimaginable. The only thing that seems solid is Dave and Jack's feelings for each other and Dave's devastation at Jack's death.[22]

––––

The plays from 1972 onward, beginning with *Small Craft Warnings,* occur, once again, in the land of the living. Their characters are sad and rueful survivors, chastened and seeking understanding and forgiveness for a multitude of sins. They will go on, harrowed but with little sense of spiritual victory. It is as if they had expected to die, thought they had died, and were surprised and perhaps a little disappointed to discover that they had not. While Williams always denied that he sought death himself, his drinking and drug-taking belied the notion that he was in love with life. If he didn't seek death, he continued to court his constant companion, flight: now, perhaps, from feelings of guilt concerning his relationship with Merlo. "We use each other and that's what we call love." He both loved and used Merlo, and accustomed as he was to blaming himself for using others, it is possible that he also blamed himself for Merlo's death. He needed forgiveness, and the plays he wrote between 1971 and 1981 suggest that he was trying to believe that he was not beyond it.

III

"Where so little was being written about us but a few years back, today the very opposite is true," wrote Edward Sagarin in 1961. Using the pseudonym Donald Webster Cory, Sagarin had published the landmark work, *The Homosexual in America,* a decade earlier, in which he affirmed that gay men need not be "cured" by psychoanalysis. "So much is being written that even the experts can hardly keep up with all the literature. Medical articles, psychological studies, technical treatises, master's theses and doctoral dissertations—hardly a subject is more openly discussed. The days of silence seem so long past that one almost forgets they ever existed." Plays were not on Cory's list, but perhaps for the first time since the *Cue* reviewer of *Garden District* had spoken of the alarming amount of gay material turning up in theatres in the 1950s, it actually appeared to be true: There were, proportionally, a large number of plays about gay characters, or with major gay characters, beginning to be produced in New York.[23]

Visibility came with a price. In the 1960s and 1970s, gay playwrights, actors, and theatregoers could not pick up a newspaper without reading, about once a month, a critic's personal view of homosexuality (almost always negative) in the guise of a review or column.

While the uptown critics rarely ventured downtown to look at the theatre that was being born, they sensed that gay voices were in the air. While the anonymous critic from *Cue* had merely commented in 1958 that, "Everyone has written of homosexuality in the last decade," three years later, this seeming overabundance had become, to many critics, a social crisis. In 1961, their dis-ease attained a critical mass, and Howard Taubman, who had recently replaced Brooks Atkinson as the *New York Times*'s lead critic, launched an assault on the front page of the Sunday Arts & Leisure section, next to the report that Tennessee Williams had grown weary of Broadway. Declaring that it was time to speak openly about "the increasing incidence and influence of homosexuality on New York's stage," Taubman accused unnamed dramatists of distorting human values by disguising homosexual characters as heterosexual ones. "Characters represent something different from what they purport to be," he wrote. "It's no wonder that they seem sicker than necessary and that the plays are more subtly disturbing than the playwright perhaps intended." Such untruthful portraiture is unhealthy, he continued, and the audience "senses rot at the drama's core."[24]

What Taubman wanted to suggest was that playwrights should be able to write openly about whatever subject matter moved them without having to resort to "heterosexual masquerade," but his attitude toward homosexuality kept getting in his way, and he never did provide a specific example of dramaturgical dissembling. Rather, he pointed to Gore Vidal's *The Best Man* (1960) and Lorey Mandel's adaptation of Allen Drury's *Advise and Consent*, both of which contain openly gay characters. He accused the playwrights of introducing homosexuality "without compelling force or overriding need." However, the examples of homosexuality that Taubman gave had nothing to do with characters per se: "It is noticeable when a male designer dresses the girls in a musical to make them unappealing and disrobes the boys so that more male skin is visible than art or illusion require." He described the strategies that gay playwrights allegedly employed to hide homosexual subtext within heterosexual characters as "insidious," "dissembling," and "furtive, leering insinuations."[25]

The attack was soon taken up by others, including the respectable academic theatre journals, and, eventually, magazines as far to the left as *Ramparts*. Disgust for homosexuals would permeate the critical atmosphere in New York well into the 1970s and even, in the case of *New York* magazine's John Simon, into the twenty-first century.[26]

What brought on these attacks? Taubman's description of yet another way one could detect hidden homosexual handiwork provides a clue: "The unpleasant female of the species is exaggerated into a fantastically consuming monster or an incredibly pathetic drab." A year earlier, a play by an unknown American playwright had opened Off-Broadway at the Provincetown Playhouse. It was simple and startling: two men chatting in Central Park, one desperately, the other politely and evasively, until, in a convulsive moment of violence, one of them kills himself. Before he does, this character, named Jerry, describes the landlady of his shabby West Side boarding house: "I don't like to use words that are too harsh in describing people. I don't like to. But the landlady is a fat, ugly, mean, stupid, unwashed, misanthropic, cheap, drunken bag of garbage." Each evening, this fantastically consuming monster and incredibly pathetic drab loiters in the entryway until Jerry comes home. There she pins him against the wall, her lemon-gin-soaked breath in his face, hoping to engage him in conversation, if not worse. Jerry also scorns the comfortable, middle-class life Peter (his new, temporary friend) lives on the Upper East Side with his wife, his pets, his two daughters, all more or less interchangeable. Jerry, by the way, announces that, for 11 days when he was 15, he was "a h-o-m-o-s-e-x-u-a-l" but now sleeps only with women. Prostitutes. "For about an hour."[27]

Like most of Edward Albee's plays, *The Zoo Story* received mixed reviews. The same was true of *The American Dream* the following year. Attacks on Albee, veiled or open, however, would thread through articles similar to Taubman's over the next several years.

A writer identified only as An Anonymous Authority, writing in *Show Business Illustrated*, in April 1962, quotes a nameless expert, described as a "best-selling author (not a playwright) who consorts with many theatre folk socially, by choice," on "the homosexual influence": "It is not easy to define this elite group within the in-group of the theatre. Theirs is a glittering world, a charming one. Conversation is epigrammatic, bitchy, brittle. Chic is dominant in their domiciles and dress. . . . I would like to stress the use of the word 'talent,' to distinguish it from genuine creativity—or genius. . . ."[28]

Albee's *Who's Afraid of Virginia Woolf?* opened at the Billy Rose Theatre on October 13, 1962. It was replete with conversation that was epigrammatic, bitchy, brittle; its author was believed to move in circles that were, at least on the surface, glittering and charming. There are no homosexuals in the play; indeed, much of the play's three hours are devoted to ruthless examinations of the characters' heterosexual failures. Nonetheless, some critics couldn't help looking for, and finding, insidious homosexual references. These critics were not restricted to the daily newspapers; even the academic journals were offering homophobia as drama criticism. In the spring 1963 issue of *The Tulane Drama Review,* the leading academic theatre journal of the era, the editor, Richard Schechner, wrote a scorching attack on Albee and the play. The play, he wrote, was phony:

> The American theatre, our theatre, is so hungry, so voracious, so corrupt, so morally blind, so perverse that *Virginia Woolf* is a success. I am outraged at a theatre and an audience that accepts as a masterpiece an insufferably long play with great pretensions that lacks intellectual size, emotional in-

sight, and dramatic electricity. I'm tired of play-long "metaphors"—such as the illusory child of *Virginia Woolf*—which are neither philosophically, psychologically nor poetically valid. I'm tired of plays that are badly plotted and turgidly written being excused by such palaver as "organic unity" or "inner form." I'm tired of morbidity and sexual perversity which are there only to titillate an impotent and homosexual theatre and audience. I'm tired of Albee.[29]

Albee's *Tiny Alice* opened at the end of 1964, and critics had another occasion to question his motives—and those of other gay playwrights. Philip Roth, in *The New York Review of Books*, took certain playwrights to task for their "dishonesty." In a review titled "The Play That Dare Not Speak Its Name," Roth called *Tiny Alice* a "homosexual daydream," in which "the celibate male is tempted and seduced by the overpowering female, only to be betrayed by the male lover and murdered by the cruel law." Roth ended by asking, "How long before a play is produced on Broadway in which the homosexual hero is presented as a homosexual, and not disguised as an *angst*-ridden priest, or an angry Negro, or an aging actress; or, worst of all, Everyman?" The angst-ridden priest was Julian in *Tiny Alice;* the angry Negro was a reference to James Baldwin's *Blues for Mr. Charlie,* which had opened on Broadway earlier that year; and the aging actress referred to The Princess Kosmonopolis in Williams's *Sweet Bird of Youth*.[30]

The following year, *The Tulane Drama Review* published a long article by Donald Kaplan, a psychiatrist, who, using *Virginia Woolf* as his principal text, laid out, at considerable length, what "homosexual drama" was, where its roots lay and why it was both depressing and injurious to public health.

In 30 dense pages, Kaplan explained that the "homosexual theatre" is rooted in infantile sexuality, where a spoiled child rebels; where he refuses to make the choice to mature and insists on having all things; where he never learns to distinguish objects outside himself from his subjective wishes or one gender from another; where he never internalizes parental authority and is always in a position of rebelling against its "No, you cannot have this," and in that regard poses as a hero and a rebel, even though his is a revolution without action (he only pouts, screams, and cries). He is, however neither hero nor rebel. He is just an egocentric infant who has not built an independent, stable identity. Kaplan argued that this is the theatre of Williams, of Inge and Albee, as well as the sexually fixated heterosexual writer such as Arthur Miller, or even Philip Roth. "The homosexual ideology," Kaplan concluded, "perpetrates the fraud of rebellion with revolution, of gain without the responsibilities of sacrifice. In current theatre, the curtain falls and returns us to a world unaltered and uninspired. The experience is humiliating."[31]

The Tulane Drama Review chronicled the groundbreaking avant-garde of The Living Theatre, The Open Theatre, and others; it published Michael Smith on the beginnings of Off Off-Broadway. The Open Theatre was begun by Joe Chaikin, a gay director; the Off Off-Broadway movement had a significant gay component. These developments could be championed in the pages of the *Review* as long as their gay side was not mentioned. Any theatre-maker, however, who was homosexual, might be subjected to virulent, unapologetic homophobia.

The "criticisms" of homosexuals in the theatre during these years centered on their supposed hatred and grotesque portrayals of women, the gratuitous display of handsome young men, and, most often, dishonesty. Gay playwrights were liars, even if, as some of the more "sympathetic" critics would write, it was not strictly their fault that they had lied, because the "public" wouldn't stand for honest portraits of gay men or women onstage. It may have been true that the "heterosexual audience" of 1961, in Taubman's words, "feel uncomfortable in the presence of truth-telling about sexual deviation." It was certainly true that the majority of critics felt this way. "If only," Taubman wrote in another Sunday *Times* piece in 1963, "we could recover our lost innocence and could believe that people on the stage are really what they are supposed to be! Would such a miracle oblige playwrights obsessed by homosexuality and its problems to define their themes clearly and honestly?" Homosexuality's "problems" included the fact that it really didn't matter whether a gay playwright wrote openly or "dishonestly." Either way, the critics would make an issue of his sexuality and a producer would be taking an even larger risk than usual producing plays that invited personal attacks in the guise of negative reviews.[32]

Some critics strived to be fair, but their pieces exposed the prejudices they thought they had overcome. In November 1965, Martin

Gottfried, the theatre critic for *Women's Wear Daily* (an important source of critical appraisal in the New York theatre when many Broadway investors were in the clothing manufacturing business), wrote a column in which he alternately insisted on tolerance and intelligence while reinforcing many of the stereotypes he was trying to condemn. "The wholesale rejection of the theatre homosexual—or any homosexual—is considered very bold," he began. "Such generalized distaste, obviously, is no more logical than any other prejudice." There are a great many homosexuals in the theatre, he continues, and that is a good thing: "The fact is that without the homosexual American creative art would be in an even sorrier state than it now is. Furthermore, there is no foundation for saying that every homosexual is 'sick'—it just is not true." Presumably, only some were.[33]

However, Gottfried felt it incumbent to describe the "terribly evident" homosexual influence on the theatre: "In choreography that gives all of the interesting dances to the men and makes the women look awkward. In costumes that are close-fitting for the boys and bulkily unattractive for the girls. In casting that is geared to the director's favorites rather than toward ability. In plays that have subtle bases in homosexual attitudes and lean toward inside jokes. In revues with Thirties flavor."

He offered another example. *Virginia Woolf,* he wrote, was "perhaps the most successful homosexual play ever produced on Broadway. If its sexual core had been evident to more people it probably never would have run—even though it is perfectly exciting theatre."[34]

By early 1966, Stanley Kauffmann replaced Howard Taubman as the *Times's* lead theatre critic. He wasted little time weighing in on the subject of disguised homosexuality and its influence in the theatre. Kauffmann was no homophobe, and his Sunday piece was clearly an attempt to place the "blame" for insidious, hidden homosexuality squarely on the prejudices of heterosexuals. It was, in this sense, a far cry from Taubman's articles. Nonetheless, Kauffmann, like Gottfried, could not help entangling himself in stereotypes he attempted neither to prove nor refute. He stated the "problem" clearly at the top of his piece, mentioning (but not mentioning) those three leading American playwrights, Williams, Inge, and Albee:

The principal complaint against homosexual dramatists is well known. Because three of the most successful American playwrights of the last twenty years are (reputed) homosexuals and because their plays often treat of women and marriage, therefore, it is said, postwar American drama presents a badly distorted picture of American women, marriage and society in general. Certainly there is substance in the charge; but is it rightly directed?"[35]

It was in his attempted defense of these and other gay writers where Kauffmann wrote himself into trouble.

If he writes of marriage and of other relationships about which he knows or cares little, it is because he has no choice but to masquerade. Both convention and the law demand it. In society the homosexual's life must be discreetly concealed. As material for drama, that life must be even more intensely concealed. If he is to write of his experience, he must invent a two-sex version of the one-sex experience that he really knows. It is we who insist on it, not he. . . .

I do not argue for increased homosexual influence in our theatre. It is precisely because I, like many others, am weary of disguised homosexual influence that I raise the matter. We have all had much more than enough of the materials so often presented by the three writers in question: the viciousness toward women, the lurid violence that seems a sublimation of social hatreds, the transvestite sexual exhibitionism that has the same sneering exploitation of its audience that every club stripper has behind her smile. (Kauffmann, 292–3)

Kauffmann assumed that because his three playwrights had, at least to his knowledge, no sexual or marital experience with women, they could neither know anything nor care about marriage or other deep relationships with them. Gay playwrights, of course, are as expert on the subject of marriage as heterosexual ones: each has had parents and each has observed and absorbed much about that particular relationship. Presumably, to Kauffmann and other straight critics, the only relationships males can have with females are that of mother or wife; friendships seem not to be possible. As for the "materials so often presented" by Kauffmann's three playwrights, it is hard to know what he means: Williams was never vicious toward women (his reputation for his treatment of women is exactly the opposite) and neither was Inge. If there is any viciousness toward women in Albee's work of the 1960s, he was no more respectful of men as a group. It is equally hard to identify "lurid violence" that is a sublimation of social hatreds. Williams was the most explicitly violent of the three playwrights, and his violence documented what he had seen of a hateful society. As for "transvestite sexual exhibitionism," this may merely be Kauffmann's way of saying he was uncomfortable seeing men portrayed as sexual beings. Many more people (especially younger ones, straight and gay) were likely to find images of Marlon Brando in a t-shirt, or Ben Gazarra and Paul Newman in pajamas, liberating.

Kauffmann's strangest charge was that gay playwrights celebrated form and style over content:

> Thus we get plays in which manner is the paramount consideration, in which the surface and mode of a work are to be taken as its whole. Its allegorical relevance (if any) is not to be anatomized, its visceral emotion (if any) need not be validated, and any judgment other than a stylistic one is considered inappropriate, even censorious. (Kauffman 293–4)

No play of Tennessee Williams could be considered all style and no substance; no one could sustain the charge that he did not at least attempt to create a visceral reaction in the audience or that his characters did not undergo them. To suggest that Inge was interested in plays of surface, style, and manner is bizarre; not one of his plays can be characterized that way. Kauffmann must be referring to Albee and his famously acid style. But to suggest that *Virginia Woolf* or *The Zoo Story* has neither content nor emotional impact beneath their style, is to suggest—certainly inaccurately—that Kauffmann had seen neither play in the theatre.

In the end, for all his liberal intentions, Kauffmann was forced to say that while heterosexuals may be at fault if homosexual playwrights feel they cannot be honest, the truth is that homosexuals hate themselves and heterosexuals, too, and their hatred cannot help but leach into their plays and infect the theatrical body. "Conventions and puritanisms in the Western world have forced them to wear masks for generations, to hate themselves, and thus to hate those who make them hate themselves. Now that they have a certain relative freedom, they vent their feelings in camouflaged form" (Kauffmann, 294).

Whatever else it did or did not do, Kauffmann's article (and, others, like Gottfried's) gave critics and audiences permission to like plays with openly gay characters so long as those characters reflected the sort of gay person Kauffmann was defining: the self-loather whose life was a misery until it ended in a slit of the wrist or a leap from a window. The only good homosexual, according to the new liberal thesis of the New York press, was the unhappy homosexual. Over the next few years, the plays that opened in New York featuring gay characters were tailor-made for just such expressions of pity and superiority.

Among the first of these was the Canadian John Herbert's *Fortune and Men's Eyes,* which opened Off-Broadway in March 1967. Its stark depiction of prison life caused a sensation.

Fortune and Men's Eyes is not about homosexuality, per se, but rather, about how sex is used as a weapon, about the way men's behavior may turn brutal when they are treated like animals. Indeed, while the production attracted attention in part because of its off-stage homosexual rape, it aroused more interest in another issue: prison reform. The producers held several panel discussions after performances, at which former prison inmates discussed the play's realism and their own, often horrific, treatment in prison. Nonetheless, the portrayal of men at the mercy of their own or others' homosexuality was just the sort of depiction that critics and the "liberal" New York audience could support while feeling good about itself at the same time. *Fortune and Men's Eyes* was a significant Off-Broadway success in the 1967–68 season, and played 382 performances.[36]

Williams was reported to be one of the play's most vocal supporters. It may be that *Fortune in Men's Eyes* struck a chord in him. He was expressing a similar view of existence in a one-act he was working on called *Confessional.* The play was a battle between two points of view. One held that life was something to be thankful for, that it could be a thing of fulfillment and beauty. The other depicted it as little more than a series of brutal encounters in which people used each other for their own purposes. Perhaps in the universe of *Fortune and Men's Eyes,* where the inhabitants were either predators or victims and sex was used as a weapon, Williams found an agreeable metaphor. Deep into his Stoned Age depression, Williams's *Confessional* is an accurate portrait of his spiritual state. The title suggests the life-long struggle between the urge to reveal and the need to conceal, with the former more explicitly prevailing for the first time. Debilitated as Williams might have been by drugs, depression, and alcohol, the struggle would possess him to the extent that he would soon turn the one-act into a full-length play.[37]

In January 1968, ten years after the story of the unrepentant Sebastian Venable opened Off-Broadway, Charles Dyer's *Staircase* debuted on Broadway at the Biltmore Theatre. Here was a play, an import from London, no less, about two gay characters whom anyone with a liberal conscience could love. Charlie and Harry, two lower-middle class London hairdressers, struggle with self-hatred. Charlie wants Harry out of the house when his daughter, whom he hasn't seen in 20 years comes to visit. To make matters worse, Charlie is about to be called up on charges of disorderliness for having been drunk and dressing in drag in public. Harry volunteers to come with him to court, a prospect that horrifies Charlie even more: "You! Come with me? God no, they'll give me ninety years." The play shows two aging homosexuals desperately clinging to a life they despise, with no prospect for happiness, joy, or even psychological comfort. In every sense, they are failures. Charlie's one claim to respectability is the fact that he was once married, and managed to have a child: "Nothing puffy with me, mate," he tells his mate. "*I'm normal. I was married with a baby.*"[38]

The critics embraced *Staircase*. Walter Kerr, now writing for the Sunday *Times*, thought it, " . . . the simplest and most honest treatment of homosexuality I have come across in the theatre." "*Staircase* is the first play that I have seen deal sensitively and adroitly with homosexuality," Gottfried agreed, adding that it was as enlightening as it was moving. Even the *Village Voice* critic, Ross Wetzsteon, was taken in. " . . . a touching, compassionate vision," he wrote, "of the emptiness and desperation beneath these glib and posturing lives." Otis Guernsey made special mention of *Staircase* in his volume on the best plays of 1968: "[Charlie's and Harry's] neurosis has destroyed everything except their continuing ability to feel and suffer. Unbeautiful as the subject of *Staircase* might sound, it is beautiful in its insight and sympathy."[39]

What pleased the critics most about *Staircase?* Its vision of the hopelessness of these two gay men's lives? Eli Wallach's mincing performance as Charlie, which drew on every physical stereotype of an old queen? The fact that Dyer suggested that gay relationships can be every bit as miserable as straight ones? Or that the characters are always haunted by the spectre of loneliness, that they are in danger of being left empty and alone, because they are gay? Donald Kaplan had made the point in his *Tulane Drama Review* article. The *Review's* house psychiatrist wrote, "Loneliness is the homosexual's most dread state of existence, the unconscious sentence of abandonment by the retaliatory parental authority; to the dependent child, abandonment is tantamount to death" (48). Donald Vining caught on quickly enough to *Staircase's* critical appeal: "The very mixed audience was very tolerant of the homosexuality," he noted in his diary, "but only [because the characters] were miserable, perhaps. I have yet to see depicted on stage or screen a homosexual couple of reasonable attractiveness who stick together even tho [*sic*] each has reasonably acceptable alternatives."[40]

A far more significant milestone of gay theatre occurred on April 14th, when *The Boys in the Band* by Mart Crowley opened Off-Broadway at Theatre Four, co-produced by Albee's producer Richard Barr. *Boys in the Band* changed the terrain for all gay drama in America and, to a large extent, the expectations gay people had of the "gay plays" they would see henceforth.

The play has acquired the reputation of being about a group of pathetic, self-hating men. In truth, only one character—Michael, the host of the party being thrown for Harold—can be described that way. (It may be that William Friedkins's 1970 film version, seen by many more people than those who saw the original Off-Broadway production, is responsible for this characterization of the play: While following Crowley's play text closely, Friedkin not only directed performances that are pathetic, claustrophobic, and verging on hysterical, he invokes a violent thunderstorm to destroy the party, as if nature itself is punishing the characters for their perverse lives.) Except for Michael, *Boys in the Band* shows a group of gay men who aren't naturally miserable because they're gay, and who actually see the possibility of happiness. Unlike the sad creatures of *Staircase* or the sadistic lesbian of Frank Marcus's *The Killing of Sister George* (which played on Broadway in 1966), the boys of *The Boys in the Band*, while certainly limited representatives of types, are not at war with their sexual natures.

The Boys in the Band has a candor that had belonged previously only to downtown theatre, even as the picture it painted of "gay life" (one not common to all gay men) wasn't entirely complimentary. Here is much bitchiness stemming from frustration and unhappiness, but no one could deny that in 1968—or at any time since—there were gay men who felt this way. However, in the effeminate Emory's refusal to "act straight" in front of Michael's heterosexual friend Alan, there is defiance of the straight world and barbed humor at its expense. Another of the boys, Hank, has left his wife and children to be with Larry, which, in 1968, was an unusual, not to say radical, act. Hank may be unhappy with Larry's unfaithfulness, but that is no more an issue of sexuality than it is in plays about straight couples who cannot rise to traditional standards of monogamy. Except for Michael, the boys seem more comfortable with their sexuality than does Alan, who beats Emery in a display of either homophobia or homosexual panic. While more than one critic quoted Michael's line, "You show me a happy homosexual and I'll show you a gay corpse," as the play's theme, *Boys in the Band* also showed gay men who could recognize self-hatred as a problem that was not of their own making and that they could overcome. Crowley himself was quoted as saying that the play wasn't about homosexuality but about self-destructiveness.[41]

There is nothing about *Boys in the Band* that the critics could point to as "disguised homosexual drama." The play told about gay life from the point of view of gay men. Its language, its attitude, even the mere fact of its existence, spoke of a certain defiance, of a long-deferred coming-out. Fundamentally, it was, as even Clive Barnes, now the daily critic at the *Times*, pointed out, a "homosexual play"—that is, a play by a homosexual, about homosexuals, occurring in a gay milieu with relatively little regard for the sensibilities of heterosexuals. This was far from the first time such a play had been produced in New York, but it was the first time one had been staged uptown and reviewed by the critics of the daily papers and national news magazines.

Critics continued to take advantage of their positions of authority to express their personal feelings about homosexuals, and interpreted the play, as they had other plays with gay characters, by emphasizing the alienation of these supposedly sad individuals. Martin Gottfried carried on his battle with himself in *Women's Wear Daily.* He laughed hysterically at *Boys'* first act, he wrote, when the humor was "wonderful and bitter fag humor." The humor was especially notable for its "added texture of desperation and irony. These imply the self-contempt and frustrations that homosexuals must fight every day of their lives—not sexual frustration but the frustration of being condemned for doing the only thing that satisfies them." Gottfried expressed sympathy for the plight of the play's characters: " . . . homosexuals have been forced into hating themselves and that so long as they do they will never be happy. I don't think that anybody would disagree with that but the pity is that knowing it doesn't change it. It must be an awful spot to be in." Only one character in the play is in that spot, but Gottfried seemed determined to view them all that way. His sympathy only went so far, however, as there are some things that just cannot be believed: "Mr. Crowley's attempts to demonstrate the real love between homosexuals are strained and unconvincing. . . . Perhaps homosexuals really can love each other but 'The Boys in the Band' doesn't show it.'" There are also some things that could not, for all one's liberal beliefs, be stomached, either: "And one more matter—perhaps it's my thing but I just can't take guys dancing with each other. It only looks like [a] pathetic imitation of men with women." The most curious aspect to Gottfried's notice is that he thought he was being positive about the play and gay men.[42]

Clive Barnes also fixed on what seemed to be the play's portrayal of homosexual angst. "The special self-dramatization and the frightening self-pity—true, I suppose of all minorities, but I think especially true of homosexuals—is all the same laid on too thick at times. . . . The power of the play . . . is the way in which it remorselessly peels away the pretensions of its characters and reveals a pessimism so uncompromising in its honesty that it becomes in itself an affirmation of life." In other words, Barnes appreciated an affirmation of gay life that recognized how really horrible it is. Otis Guernsey also felt the piece positively uplifting in its depiction of lost souls, and was happy to observe that Robert Moore's direction produced "a harrowing portrait of the neurosis itself," while "giving no offence to taste or decency."[43]

Other critics, including George Oppenheimer of *Newsday*, who described Michael's friends as "somewhat less than all-male guests," tried their best to like the play but criticized presumably gay members of the audience for their unrestrained response to the sexual humor—which they undoubtedly saw in more instances than the straight critics did (Oppenheimer referred to their "shrill shrieks"). These critics failed to understand the fundamental event that *Boys in the Band* was for its gay audiences: A community was gathering to see itself portrayed onstage for the first time in such a highly visible, uptown venue with something approaching accuracy, told from a gay point of view and, in doing so, they recognized aspects of their own lives. Heterosexual audiences take characters' heterosexuality for granted and have always been allowed to see beyond it to other aspects of their lives. However, expressions of sexuality were among the first things in *The Boys in the Band* to have a visceral affect on gay audiences unused to seeing people like themselves represented on a stage. Their enthusiastically vocal responses to Emory's multiple double-entrendres was a mixture of appreciation and recognition, a great "Yes" to what they were seeing—finally—on a stage.[44]

Only Michael Smith in his *The Village Voice* review found it unnecessary to state his personal views on homosexuality. He took issue with what he presumed might have been self-loathing, and pointed out that Crowley's play did not represent the lives of every homosexual: "The play's faults are its pretension to being about homosexuals in general, rather than a particular gay social scene; its implication that these general human miseries are specific symptoms of homosexuality; and its tendency to explain and preach."[45]

Donald Vining's diary entry about *Boys in the Band* is revealing for the way it demonstrates that a gay man in the audience did not necessarily notice the play's so-called self-hatred as much as its optimism, and also for its observation that any number of gay men shared Michael's ambivalences while still leading happy lives:

In the first act we screamed with laughter as the gay party got under way but the second act, as they got drunker and nastier, was much more sober. Fortunately, only one of the characters was terribly campy, one was mildly so, and the rest were allowed some dignity and masculinity despite their sexual proclivities. The scene where they played a game in which each man phones the one person he loved best to tell him so was a bit unbelievable but very well done and touching. At least one couple was allowed some happiness together and despite jealous scenes, chose to call each other during the game and eventually retired to the bedroom to make love.[46]

He also noted that the audience was much more a mixture of gay and straight couples than it had been at *Fortune and Men's Eyes*.

Boys in the Band played 1,001 performances. Tickets for seats in the first seven rows at the 299-seat Theatre Four on West 54th Street that cost $5.95 were being scalped for $25, and the play recouped its investment in 11 days. Clearly, gay and straight New Yorkers hungered to see this play.[47]

IV

What would be the implications of *Boys in the Band* for Tennessee Williams? It is Harold who spells things out for Michael in the play's most famous speech: Michael is a homosexual and, however much he'd like to change that fact, he cannot, not with all the prayers to his Catholic god or all the psychoanalysis he can charge on a credit card. While some may regard Harold's speech as a mere scolding to a self-loather, it is more correctly viewed as a wake-up call to self-acceptance. Tennessee Williams's plays with gay characters might make a similar call indirectly, and would subtly contrast a gay character's unhappiness with that of the straight characters', as he did in *Confessional* and would again in *Small Craft Warnings*. But in the din of celebration and the demands for social and political equality aroused by *Boys in the Band*, the Stonewall Riots, and the emergence of gay liberation, subtle distinctions stood little chance of being heard. *Boys* was not the kind of play Tennessee Williams ever had written or ever would, but this was in part a result of political culture and imperatives. Of all the plays Williams wrote after 1968 that placed gay characters at the center of their stories only one, perhaps, could be viewed as a gay play: *Something Cloudy Something Clear*, in which August's feelings for Kip, and his guilt for what he considered his exploitation of him, make up the play's principal material. After *Boys in the Band*, many in the gay audience, if not among the straight critics, grew more ready for plays in any genre about gay *life*, not those that merely included gay characters.

The years of abusive treatment of homosexuals that newspapers, magazines, and journals routinely dispensed under the guise of criticism also would have its effect on Williams's reputation among young gay men and women. Just as they were becoming fed up with their treatment by police and politicians, at least some gay and lesbian theatre-makers were becoming sick and tired of the homophobia they regularly encountered in the cultural press. The defiant downtown theatre they were making was one response. That Williams would not publicly answer critical insults for many years did not help. Of what use, what relevance to their lives was a gay playwright who wrote by indirection, was even-handed in his treatment of heterosexual and homosexual characters, who did not use his fame to speak out and be "proud" in an era that would increasingly insist on pride and political correctness as artistic criteria?

Three days before *The Boys in the Band* opened, Donald Vining noted in his diary that a "Student Homophile League" had been founded at Columbia University where he was now employed, and that its founder, a junior named Stephen Donaldson, had written a long article in the student daily, *The Spectator*, called "The Anguish of the Homosexual Student." Vining remarks on a term Donaldson used that he himself had never heard before: "in the closet."[48]

Perhaps it was the Stonewall uprising in June 1969, and the bursting on the scene of gay liberation that caused Williams to speak publicly for the first time, outside the world of his plays, about his homosexuality. Unsurprisingly, he seemed to be ambivalent when he came out to the general public. He made a voluntary, if not exactly specific, remark in reply to a question by David Frost on Frost's interview program, January 21, 1970. In an *Atlantic Monthly* interview written by Tom Buckley in November of that year, he talked freely of the probably fictitious "Green Eyes," who crawled into his fraternity house bed during his student days at the University of Missouri.[49]

At about the same time, he was interviewed in New Orleans by Rex Reed for *Esquire*. He complained that the Buckley piece had provoked episodes of violence against him and his property in Key West; that youths began throwing eggs and rocks at his house, and doctors refused to treat him for a fall he took at night. Reed asked him if the theatre wasn't a safe place for gay people to work. Williams agreed that people who actually worked in the theatre were broad-minded and understanding, but that the critics were less so. Finally, he responded to the charges made in the 1960s about gay writers disguising homosexual relationships as sexual ones.

> I've read things that say that Blanche was a drag queen. Blanche DuBois, ya know . . . that George and Martha in *Who's Afraid of Virginia Woolf* by Albee were a pair of homosexuals . . . these charges are ridiculous! . . . If I am writing a female character, goddamnit, I'm gonna write a female character, I'm not gonna write a drag queen! If I wanna write a drag queen, I'll write a drag queen . . . [50]

It wasn't Williams's growing paranoia that prompted him to complain about a sudden feeling of being unwelcome in Key West. David Lobdell, a young Canadian poet whom Williams had invited to watch his Key West house in the fall of 1970 while he and his friend Oliver Evans visited Japan and Thailand, observed this firsthand. Lobdell wrote his sister that December,

> There are a great many people in this town who don't like Tenn, who feel that he is a disgrace to the town, and who take it upon themselves to "punish" him, and whoever happens to be associated with him; because I am living in the house, I am guilty by association. Eggs are thrown at the house at night, and junk is tossed over the fence into the pool; people phone late at night and shout obscenities into the mouthpiece . . ."[51]

Even in Key West in 1970, a famous playwright was not safe after coming out.

Some gay activists and writers were not impressed by Williams's public statements or the supposed lack of gay characters in his work. One of them, writing under the pseudonym "Lee Barton," wrote a play called *Nightride,* which opened Off Off-Broadway in December 1971. The play lays out an intergenerational conflict between a famous closeted gay

playwright named Jon Bristow and a young rock star, Jab Humble, who
has recently come out in front of a concert audience. Bristow has written
few good, or even produceable, plays since the death of his lover years
earlier, and hasn't had a hit in a decade. He has spent most of the ensuing
time drinking and taking pills, and now lives in Puerto Rico with a new
lover, a young sailor named Erik. Humble wants to buy the rights to a
book of homoerotic poetry Bristow wrote in 1939 when he was 18 called
The Kama Sutra on Third Avenue and use the verses as lyrics. Bristow is
appalled—selling Humble the rights and making the poetry public
would mean coming out, and that, he believes, would ruin his name and
his career. Bristow's grasping agent urges him to make the deal: The
young audience not only thinks he's out of touch with contemporary life,
they laugh when Bristow is referred to as a living writer.

Humble arrives in person to plead his case, but Bristow, terrified of
being exposed, refuses the quarter-million dollars Humble offers him.
The chasm between them seems unbreachable. Humble refers to the gay
characters in Bristow's plays as sick; Bristow is amazed that Humble is actually proud to be out and gay. Humble insults him not only as a drunk
who can no longer write, but as a coward, as well. In the end, however,
after Erik intercedes on Humble's behalf, Bristow gives the young rock
star the rights to the poems at no charge and at last confesses that he is
tired of being afraid.

Now on the wagon, the playwright describes his long years of falling
down (as Williams was notorious for doing during the 1960s). He admits
that he disguised the great gay love of his life as a woman in a hit play
staged in 1953, but defends himself against Humble's charges of homophobia and exploitation by describing the way things were in the 1940s
and 1950s, when stage and film careers could be destroyed if one's private
gay life was discovered, about how audiences wouldn't stand for anything
truthful about homosexuals.

It required no special knowledge for audiences to recognize the inspiration behind Jon Bristow. One critic, at least, identified Williams as the
playwright behind Bristow:

> The central figure of Lee Barton's play *Nightride* is an aging homosexual
> playwright who has not written anything worthwhile in 10 years. He has

been on a collision course aimed at Skid Row, occasioned by the loss, many years earlier, of his true love, a sensitive and compassionate youth. The entire action of the play takes place in the Puerto Rico beach-house which the playwright and his current lover—a sailor—occupy. Into this glass menagerie enter the playwright's literary agent . . . and another homosexual couple: a hip-swinging rock music star who wants the rights to explicit love poems written many years earlier by the playwright, and the mute lover of the musician.[52]

In an article published in the Sunday *Times* in January 1972, "Barton" defended himself against charges of hypocrisy for using a pseudonym while accusing another gay writer of cowardice. "Barton" wrote that only the wealthy and famous, or the socially outcast, could afford to come out in 1972. People like himself, unknown, powerless, and the holder of "a regular desk job in industry" risked dismissal and even the investigation of his friends if his homosexuality was a public fact. This is just one reason why artists like Tennessee Williams owed it to gay people to write positive images of gay men.

> One work of art dealing truthfully with homosexual life is worth a hundred breast-beating personal confessions. Who really gives a damn that Tennessee Williams has finally admitted his sexual preferences in print? He has yet to contribute any work of understanding to gay theatre, and with his enormous talent one of his works would indeed be worth any amount of personal data. And several others of his generation of writers, as well as some younger ones, all of them gay, have failed to come forth with anything, under *any* name, that would make a valid case for the homosexual in society. In fact, if anything, their concealment has been so gross and so un-thought out that some critics have called it repeatedly to the public's attention.[53]

This time, Williams did not wait six years to respond. He told Arthur Bell of *The Village Voice,* "I feel sorry for the author. He makes the mistake of thinking I've concealed something in my life because he writes under a pseudonym. I've nothing to conceal. Homosexuality isn't the theme of my plays. They're about all human relationships. I've never faked it."[54]

Williams saw *Nightride* while holding auditions for his next play. In the new work, *Small Craft Warnings,* he would be far more direct about life as a gay man than he had been before. If, however, gay audiences were going to demand plays that simply valorized their lives, or that made a conscious effort to stand up for gay people rather than suggest that as human beings their problems and loneliness were no more easily resolved than those of heterosexuals, they would be disappointed.

V

The Stonewall Riots, on the weekend of June 27–28, 1969, ignited the spark that flamed into the modern gay liberation movement. Williams, whatever his spiritual or physical condition, closely followed it as well as women's liberation and the anti-war movement and supported the young people engaged in them. But he did not seek to influence events through his writing. Most often, his plays from this period, with the exception of *The Red Devil Battery Sign,* did not even reflect them. While denouncing the Vietnam War and the oppression of African Americans, as he did in a 1966 interview with Walter Wager, he said he would not write about these subjects directly—at least, no more directly than he had in a play like *Orpheus Descending.* "I am not a direct writer," he said.

> I am always an oblique writer, if I can be; I want to be allusive; I don't want to be one of these people who hit the nail on the head all the time. . . . I'm not a person dedicated primarily to bettering social conditions, because I am not able to, except through my writing, and I doubt whether people will pay enough attention to writing for writing to have any effect.[55]

By the 1970s, a second generation of gay theatre had emerged, even more distant than its 1960s predecessor from Williams's sensibilities and concerns. This new theatre was forged in the political consciousness of the Stonewall Riots and liberation. A great many of its playwrights and directors believed that people *would* pay attention to writing, and that the writing *could* make a difference in their lives. Just as new, radical political organizations, such as the Gay Activists Alliance and the Gay Liberation Front, arose in New York to take the place of the by-now staid Mattachine Society, so new theatres were founded to produce specifically gay plays for gay and lesbian audiences. Some of the plays these theatres produced were about the act of coming out, but many more took being out for granted. Cino alumnus Doric Wilson founded TOSOS (standing for The Other Side of Silence) in 1972; in 1976, John Glines (who would later produce *Torch Song Trilogy* and *As Is*) established The Glines, a company that produced gay plays well into the 1990s. There · was also the Stonewall Rep, the Meridian Theatre, the Shandol Theatre, Medusa's Revenge, More Fire Productions, It's All Right to be Women, and the Womanrite Theatre Ensemble. In 1977, the Gay Men's Theatre Collective and Theater Rhinoceros began producing plays in San Francisco.

The gay playwrights who were writing in these years would create work that was gay in content, style or both. The gay theatres were producing gay plays for gay audiences; nothing had to be concealed, and the mere revelation of a character's sexuality was largely passé. These gay theatres of the 1970s had a significantly different agenda than their Off Off-Broadway predecessors. The earlier gay plays, by and large, were the early explosions of authentic self-expression by gay writers who, for the first time, had a forum and a form for telling truths about their own lives. Most did not write from an overtly political agenda; rather, they were re-ordering the events and inner experiences of their lives in a coherent, artistic form. That the beginnings of a modern gay identity and community were emerging from this work was a by-product of the plays produced on the stages of the Cino, the Judson Poets' Theatre, and La Mama, not their reason for being.

The plays produced by the theatres that identified themselves and their audiences as gay in the years between Stonewall and the advent of AIDS, were a different matter. They were (many intentionally, some not) political. After Stonewall, at least in cities like New York and San Francisco, there was little that separated personal and sexual statements from political ones. Irrespective of their immediate subject matter, the plays produced in gay theatres between 1969 and 1983 were political acts. In those years, when gay and lesbian activists were consciously engaged in the business of identity- and community-building, any play produced by a gay theatre for a gay audience, no matter how trivial its content, became a statement about being gay. Holding hands with your lover in the theatre was a political act.

No matter how much he might have liked to be considered hip and up-to-date (which meant an unequivocal embrace of gay liberation), Williams could write only what he could write, what he had to write. *Small Craft Warnings* exists in the psychic territory between two poles, neither of which had much connection with the political and aesthetic changes experienced and often brought about by younger gay women and men. One pole is exemplified by the sentiments he put in a letter to Kenneth Tynan as early as 1955:

> In my case, I think my work is good in exact ratio to the degree of emotional tension which is released in it. In a sense, writing of this kind (lyric?) is a losing game, for steadily life takes away from you, bit by bit, step by step, the quality of fresh involvement, new, startling reactions to experience, the emotional reservoir only rarely replenished [. . .] and most of the time you are just "paying out," draining off. [. . .] Sometimes the heart dies deliberately, to avoid further pain. [. . .] Once the heart is thoroughly insulated, it's also dead. At least for my kind of writer. My problem is to live with it, and to keep it alive.[56]

The other pole is expressed in another letter, this one to David Lobdell. In September 1968, Williams wrote Lobdell, who had expressed depression and unhappiness to Williams:

> David, despondency passes, at least to some degree. You have to endure it with patience for the while that it stays. In so many letters of yours there has been a feeling of—what?—life's morning. . . .
> I offer you a ghostly hand, David, on your hand, your face, or your hair. Remember your priceless owning of life's morning![57]

Small Craft Warnings is a chronicle of Williams's Stoned Age, his half-life in the 1960s, when the possibility of life's morning was beyond his reach, when he had deadened his spirit and heart with drugs and alcohol to avoid further pain. Gay critics have found the play to be irredeemably, even virulently, heterosexist; Nicholas de Jongh says that *Small Craft Warnings* "defines all homosexuals in terms that the homophobes of the 1950s had made their accusatory own." This is one way, of course, of viewing *Small Craft Warnings*. It is a narrow view, arrived at by a selective scrutinizing of only one of the play's nine characters. The play has nothing to do with homophobia, really; the problem shared by all the characters is a phobia for feeling, and for living.[58]

The play takes place in a bar called Monk's Place on the Southern California coast. This bar, however, cannot be mistaken for the grimly realistic one of, say, *The Iceman Cometh*. Williams specifies that, "*Ideally, the walls of the bar, on all three sides, should have the effect of fog rolling in from the ocean.*" Further, as the lights come up, we hear the sound of ocean wind. Above the bar is the set's principal visual element: "*[. . .] over [the bar] is suspended a large varnished sailfish, whose gaping bill and goggle-eyes give it a constant look of amazement.*" The characters who inhabit this place are not merely living at the bottom of a bottle; they are at the bottom of the sea, vessels sunk in a sea of disappointment and enveloping emotional numbness. Indeed, in a sense, this play is Williams's *Iceman Cometh*, where he laments the loss, not of pipe dreams, but of the welcoming, questing, living spirit. The critics who find little in the play but homophobia don't notice that every character in it is wandering through the emotional barrenness of the playwright's Stoned Age, in which Williams had given up on practically everything but his means of escape from life's pain: alcohol, pills, injections, and, every morning, writing.[59]

Life's morning could not be further off for these characters. It is evening when the play begins and deep night when it ends, far from that moment when life seems born again. One can often determine a great deal about a play's themes according to the first thing we see happen, and in *Small Craft Warnings*, the first thing to occur is Monk, the bartender and owner, noticing that Violet, one of his regulars, has brought her battered suitcase with her. To Monk, this means she is hoping to move in, or at least spend the next several nights with him in his living quarters above the bar. Monk is of two minds: "I'm running a tavern that's licensed to dispense spirits," he says to Doc, another regular who is sitting at the bar, "not a pad for vagrants" (226). On the other hand, Monk, like the other denizens of his place, is lonely and aching for companionship, and, like Williams, is afraid of dying alone in the night. Among other things, *Small Craft Warnings* is about the conflicts that arise between two people when their inextinguishable desires for love, sex, and companionship are thwarted by fear of heartbreak and pain. The play suggests that one strategy we employ to avoid the pain and damage is to flee physically; other tactics include alcohol and the cultivating of an exterior so thick and resistant that it insulates us not only from hurt but also from feelings of any kind. In this sense, the play is revisiting themes Williams had been writing about since *Battle of Angels.*

Gay critics have focused almost entirely on Quentin, the homosexual down-on-his-luck, hack screenwriter. Our first glance of him, when he enters the bar with a young boy he has picked up, would seem to confirm the notion that the play is "heterosexist." Certainly, by 1972, Quentin's appearance, as described by Williams, was not one of a young contemporary gay man walking the streets of New York on his way to a meeting of the Gay Liberation Front: "*[Quentin] is dressed effetely in a yachting jacket, maroon linen slacks, and a silk neck-scarf*" (240). Williams, comments John Clum, "has returned us to the stereotype of the homosexual as fop."[60]

Critics frequently categorize Quentin as a self-hating homosexual by excerpting a speech in which he describes a sad life of gay promiscuity. Here is how one critic quotes it:

> There's a coarseness, a deadening coarseness, in the experience of most homosexuals. The experiences are quick, and hard, and brutal, and the pattern of them is practically unchanging. Their act of love is like the jabbing of a hypodermic needle to which they're addicted but which is more and more empty of real interest and surprise. This lack of variation and surprise in their . . . "love life" . . . [*He smiles harshly*] . . . spreads into other areas of . . . "sensibility?"[61]

Taken by itself, the speech sounds self-hating, indeed. But when Quentin and this speech are viewed as threads in the fabric of the entire play, a different impression emerges. The above excerpt comprises four sentences in a monologue that is a page and three-quarters long. The lengthy portion that is excluded contains the germ of Quentin's unhappiness—and that of the unhappiness of all the play's characters. It is important to quote it at length:

> Yes, once, quite a long while ago, I was often startled by the sense of being alive, of being *myself, living!* Present on earth, in the flesh, for some completely mysterious reason, a single, separate, intensely conscious being,

myself: living! [. . .] Whenever I would feel this . . . *feeling,* this . . . shock of . . . what? . . . self-realization? . . . I would be stunned, I would be thunderstruck by it. And by the existence of everything that exists, I'd be lightning-struck with astonishment . . . it would do more than astound me, it would give me a feeling of panic, the sudden sense of . . . I suppose it was like an epileptic seizure, except that I didn't fall to the ground in convulsions; no, I'd be more apt to try to lose myself in a crowd on a street until the seizure was finished. . . . They were dangerous seizures. One time I drove into the mountains and smashed the car into a tree, and I'm not sure if I *meant* to do that, or . . . In a forest you'll sometimes see a giant tree, several hundred years old, that's scarred, that's blazed by lightning, and the wound is almost obscured by the obstinately still living and growing bark. I wonder if such a tree has learned the same lesson that I have, not to feel astonishment any more but just go on, continue for two or three hundred years more? . . . This boy I picked up tonight, the kid from the tall corn country, still has the capacity for being surprised by what he sees, hears and feels in this kingdom of earth. All the way up the canyon to my place, he kept saying, *I can't believe it, I'm here, I've come to the Pacific, the world's greatest ocean!* . . . as if nobody, Magellan or Balboa or even the Indians had ever seen it before him; yes, like he'd discovered this ocean, the largest on earth, and so now, because he'd found it himself, it existed, now, for the first time, never before. . . . And this excitement of his reminded me of my having lost the ability to say: "My God!" instead of just: "Oh, well." (260–1)

Williams is measuring the distance Quentin has traveled from the ability to be surprised by life to barely having the energy to summon up an "Oh, well." He also drastically telescopes the time it takes him to move from one extreme to the other—Quentin is still a young man. It had taken Williams the better part of his life to travel the same deadening ground.

It is not being a homosexual that Quentin or Williams mourns here, *it is having lost the capacity for surprise,* and it is a feeling that the other characters, with one exception, share. For all of these characters, heterosexual and homosexual, mere sex, automatic, unsurprising, like the jabbing of a hypodermic needle, has become an inadequate substitute for spontaneous awareness. They circle about each other, alternating the roles of moth and flame, seeking out comfort and warmth even if it is only in the form of loveless sex: Doc, who has lost his medical license due to heavy drinking (but who still practices on the sly); Violet, a youngish lost soul unable to cope with the world and something of a nymphomaniac; Bill, a self-defined stud who lives by his cock; Leona, perhaps the strongest of the lot, an earthy hairdresser who accepts everything life offers except cruelty and loneliness; and Steve, a feckless middle-aged man stuck in a dead-end job as a short-order cook.

Precisely how inadequate a substitute sex is for the heterosexual characters is clear. Violet always has her hands under a table to feel the erection of whichever man is nearest. Leona has taken in Bill, although he is callous and cruel, to have his warm body next to hers. Whereas Quentin *speaks* about the negative aspects of some gay sex, Williams *dramatizes* the negative aspects of some straight sex throughout the play. The relation between sex and love, even between sex and pleasure experienced by these characters, is laid out by Doc five minutes into the play. He describes to Monk his most recent conquest:

> Remember that plump little Chicano woman used to come in here with me some nights last summer? A little wet-leg woman, nice boobs on her and a national monument for an ass? Well, she came to me for medical attention [. . .] She had worms, diet of rotten beef tacos, I reckon, or tamales or something. I diagnosed it correctly. I gave her the little bottle and the wooden spoon and I said to her, "Bring me in a sample of your stool for lab analysis." She didn't know what I meant. Language barrier. I finally said, "Senorita, bring me a little piece of your shit in the bottle tomorrow." [. . .] Some beginning of some romance. (226–7)

This tender story is immediately followed by Bill's entrance. This is how Williams describes him: "*Bill enters the bar; he comes up to it with an over-relaxed amiability like a loser putting up a bold front: by definition, a 'stud'—but what are definitions?*" (227). Bill puts more emphasis on his sexual prowess than would a man who really believes he has it. "You got arms on you big as the sides of a ham," Violet says to him with admiration. "That ain't all I got big," he replies (229). He invites her to grope him beneath the table; she accepts the offer. This tawdry kind of sex seems to be the one thing they can do in order to feel alive, and to be noticed by others.

Like many straight men who are insecure about or uncertain of their sexuality, Bill is a homophobe. He can't hear of Leona's late gay brother Haley without mocking him: "She had this brother, a faggot that played the fiddle in church, and whenever she's drunk, she starts to cry up a storm about this little fag that she admitted was arrested for loitering in the Greyhound bus station's men's room, and I say, 'Well, he was asking for it,' she throws something at me" (229). When Quentin and Bobby enter, he observes, "Y' can't insult 'em, there's no way to bring 'em down except to beat 'em and roll 'em" (241). This is followed by a bragging monologue in which Bill predicts that Quentin will soon follow him into the lavatory, and then arrange to meet him later. At that rendezvous, Bill will threaten Quentin and rob him. He does, however, make a point of telling us that he doesn't like to beat up queers: "They can't help the way they are. Who can?" (242).

Directly after this monologue comes Steve's. None of the critics concerned with the play's alleged homophobia notice the concentrated self-hatred of this speech, which expresses a despair and loathing far beyond anything spoken by Quentin. Steve pursues Violet because he figures he doesn't deserve any better than this poor, nymphomaniacal young woman.

> I guess Violet's a pig, all right, and I ought to be ashamed to go around with her. But a man unmarried, forty-seven years old, employed as a short-order cook at a salary he can barely get by on alone, he can't be choosy. Nope, he has to be satisfied with the Goddam scraps in this world, and Violet's one of those scraps. She's a pitiful scrap, but . . . [. . .] something's better than nothing and I had nothing before I took up with her. [. . .] Oh, my life, my miserable, cheap life! It's like a bone thrown to a dog! I'm the dog, she's the bone. (242)

These self-hating heterosexuals engage in a large dose of self-pity. Quentin does not.

The third gay character in *Small Craft Warnings* is Haley, Leona's brother who died at an early age from pernicious anemia. He was a violinist whose playing was so angelic that when he performed in church services, even the stingiest would put money in the collection plates. "Haley had the gift of making people's emotions uplifted, superior to them!" (247). Haley is heavily sentimentalized (and to that extent, dehumanized) by Leona, but to her he represents the "one beautiful thing" that people like Bill lack, the one thing that saves the heart from corruption. What is that one thing? Some critics look at Haley and can see only another dead, offstage gay character. This apparent self-loathing of Williams's is only amplified by Leona's description of "the gay scene" that she learned from Haley, who came out at an age younger than Bobby's: "I know how full it is of sickness and sadness; it's so full of sadness and sickness, I could almost be glad that my little brother died before he had time to be infected with all that sadness and sickness in the heart of a gay boy" (254).

In an interview that appeared in *The Saturday Review* shortly after *Small Craft Warnings* opened, Williams told the interviewer that Leona's long monologue memorializing Haley, who passed through the world with the swiftness and unearthly beauty of a comet, was the central scene of the play, the image that came to his mind first, and around which he built the rest of the play when it was still the one-act *Confessional.* What is central about that speech, and about the image of the late Haley, is not the homosexuality that he and Quentin share; it is the innocence, the lack of corruption, that they do not. Haley is that beautiful thing Leona holds sacred, the innocent who, saved from corruption by death, never had the opportunity to use others to further his own desires or be used by others or be victimized by their homophobia.

Who, among this collection of human driftwood is content and well-adjusted? It is Bobby, the young man picked up by Quentin. John Clum tells us, "Bobby is not homosexual. He is healthily, polymorphously perverse. Like Joe Orton, Williams could romanticize the possibility of such carefree bisexuality but could not present a positive picture of homosexuality."[62] Indeed, Bobby has slept with women—well, with *a* woman—but this does not disqualify him as a homosexual, and in any case, his sexuality is beside the point.

Bobby, alone, has not yet lost the capacity for amazement. An Iowa boy who has ridden his bicycle cross-country on his way to Mexico, he can be surprised and delighted by things the others cannot: "I guess to you people who live here it's just an old thing you're used to, I mean the ocean out there, the Pacific, it's not an *experience* to you any more like it is to me. You say it's the Pacific, but me, I say THE PACIFIC!" To which Quentin ruefully replies, "Well, everything is in 'caps' at your age, Bobby" (256). Nor does Bobby judge Quentin the way Bill does. He says to Leona, "That man didn't come on heavy. His hand on my knee was just a human touch and it seemed natural to me to return it" (263).

Most of all, Bobby recalls a young Tennessee Williams who, in 1939 at the age of 28, at last in full possession of the knowledge of his sexuality, set out for California from New Orleans with his friend Jim Parrott. For a time, he worked on a squab farm outside Los Angeles, and rode his bicycle back and forth from the city. He pedaled down to Laguna Beach and to Tijuana. Williams later recalled these days as some of the most carefree of his life.[63]

The scenes between Bobby and Quentin mark the distance between the young Tom Williams who went out to California and rode his bike down to Mexico, feeling full of life and possibilities, and the Williams of 1973, after Frank, pills, alcohol, and loneliness. The old Tennessee Williams confronts the young Tom, who is too fresh, too innocent, too inexperienced, too fortunate to recognize himself.

If there is in Williams some of Quentin and Bobby, Leona is a part of him, as well. Leona is the most open-hearted of the group; she has never, she says, said no to life. She befriends the otherwise-friendless Violet, bringing her not only food when she is destitute and ill, but offering plates and silverware, as well. When she feels betrayed by Violet and Bill, she explodes in a violent rage commensurate with her previous feelings of friendship. She is the only one in Monk's Place who has not dulled her feelings or obliterated a strong sense of humanity with drink. She is sober enough, usually, to long to care for someone and to be cared for; what she fears most of all is loneliness. Leona has paid a price for insisting on remaining a vulnerable being. She is bound to be let down; she lives, at least in part, in the expectation of being let down again. There is a reason why she has made something of a fetish out of her late brother: he is perfection, the one perfect and beautiful thing she has known, and he will never let her down. Like Williams, Leona cannot remain in one place for long, because she knows she will inevitably be betrayed by someone she lets into her life, as she has been by Bill, who regularly steals cash from her purse and then, on this particular night—the anniversary of the sainted Haley's death—fails to come home for the special dinner Leona prepared. Instead, Leona finds him in the bar with Violet, who is masturbating him beneath a table.

So at the end of the play she will pull up stakes once again, and move on to a new town. She intends to be a "faggot's moll," to find a young gay man who needs a friend. This would not seem to necessarily be the best way to stave off her greatest fear, loneliness. But then, neither does living in a mobile home, which she can easily pack up and drive to the next place where she will be alone again until, like Tom Wingfield, she finds companions in the nearest bar. Loneliness scares and unites all the bar's inhabitants, with the exception of Bobby. Doc drinks to escape it; Violet looks compulsively for sex; even the laconic Monk is not immune.

In his drunkenness, Doc allows a woman to hemorrhage to death following the birth of her stillborn child. Given to philosophizing when he is in his cups, which is all the time, he is much like Dr. Chebutykin of *Three Sisters*, and we are reminded of how important an influence Chekhov was on Williams (as we are when we hear Steve's sad wail, "Oh, my life, my miserable, cheap life!") Like Chekhov's characters, the patrons of Monk's Place want very much to live, to be surprised, but they have forgotten how. Like Chekhov, Williams tries not to judge them. A radio announcer warns, "Heavy seas from Point Conception south to the Mexican border, fog continuing til tomorrow noon, extreme caution should be observed on all highways along this section of the coastline." Monk remarks, "Small craft warnings, Doc." Doc replies, "That's right, Monk, and you're running a place of refuge for vulnerable human vessels [. . .]" (229).

It may be that Williams did not succeed as well as Chekhov did in withholding judgment, but in the world of his plays heterosexuals and homosexuals are equally unhappy. That is at least even-handed—and it suggests that the question of sexuality does not enter into Williams's reality of loneliness at all.

Loneliness, an incapacity for surprise; these and a third trait unite the characters of *Small Craft Warnings*. The knowledge that we use others for our own ends, and the guilt that comes with that knowledge, is present, as well. All the characters use each other to assuage their loneliness; even Monk, who gives these forlorn people a place to gather and ease their pain with alcohol, is in the exploitation business: he turns their terrible loneliness into profit. In this play, however, and in most of those to follow, there is little struggle or protest against this universal condition. Too tired, perhaps, to whip himself into the baroque objections of *Suddenly Last Summer,* in *Small Craft Warnings* Williams rues what seems now to be an inevitable fact of life. It is another, sadder cry of *En avant!*

━━━━━━━━━

Certainly, Williams's Stoned Age had taken a severe toll on his work. A stage direction at the opening of the play is indicative of the extent to which his creative energy had been depleted: *At some time in the course of the play, when a character disengages himself from the group to speak as if to himself, the light in the bar should dim, and a special spot should illuminate each actor as he speaks* (225). There are a great many monologues in *Small Craft Warnings,* and most contain material that Williams once would have dramatized. The long monologues in *Suddenly Last Summer* are narratives but they are dramatic, as well: The stakes for the speakers are high, and they need their stories to affect their listeners, and the speeches often change the course of the action. The monologues in *Small Craft Warnings* are purely narrative: They are addressed, dramatically speaking, to no one, and nothing comes of them. They are lyrical elegies, often moving in themselves. While our understanding of the characters may be advanced by them, the action of the play, the movement of the conflict, is not. Williams's energy to create monologues that were dramatic and not merely narrative, let alone to create action, had gone. In the 1960s, Williams's world had shrunk to the smallest dimensions. He had been speaking to himself in those years, and perhaps, as far as the gay audience was concerned, he still was. "Monk's Place" is a good description of where Williams had been living, spiritually, in his Stoned Age: in a lonely monastery of alcohol, drugs and oblivion, a self-imposed exile. Gay audiences were likely to have been at best bemused and at worst scornful of some of Williams's language that suggested that he was out of touch with gay people; the imagery of Quentin and Haley might have struck audiences as quaint as Bobby's term for his bicycle "my speed iron"), or Leona's reference to becoming a "faggot's moll." For exuberance (and for many gay men, the early 1970s were nothing if not exuberant) Quentin's "Barman? . . . Barman? . . . What's necessary to get the barman's attention here, I wonder?" (241) hardly compared to Emory's famous, "Who do you have to fuck to get a drink around here?" in *The Boys in the Band* (22). At a time when many gay men were in the midst of a much-delayed celebration of self, Williams produced a picture of his own life at its loneliest and most desolate.

By the time Williams emerged from his Stoned Age many in gay and lesbian theatre would be impatient, if not enraged, by characters such as Quentin who seemed, at first glance, to be pathetic, self-loathing creatures from an earlier era. Now, gay and lesbian playwrights were writing about being "out and proud," about liberation, about living one's life as a political act. This was not an atmosphere in which Williams, for all his sympathy with gay liberation, could thrive. He was simply not an overtly political writer; his writing demanded a listening between the lines, and a recognition, in *Small Craft Warnings,* that the despair Quentin feels is a human despair felt by the play's heterosexual characters, as well.

The critics did not help. They were unable to see six lonely heterosexuals if there was one lonely homosexual among them. In the *New York Post,* Richard Watts thought that Quentin and Bobby were homosexuals as "grim" as those depicted in *Boys in the Band:* "The play neither mocks nor romanticizes them, and both are seen as lonely figures who are having no easy time of it." He did not seem to sense that the straight characters are equally lonely, more desperate. Harold Hobson, writing of the London production for London's *Sunday Times,* also could not see past sexuality to the true roots of loneliness: " . . . the homosexual, in a curious and interesting speech, meditates quietly on the imaginative barrenness that follows sexual perversion." In a sympathetic review in *Time* magazine, Ted Kalem characterized Quentin as "sardonically nihilistic," and in a news story in *The Daily News,* Tom McMorrow referred to him as a "tormented homosexual."[64]

However, the play did well enough to move from the small Truck and Warehouse Theatre to the larger New Theatre in June, but it required publicity. Williams did dozens of interviews; he delivered a weather report on a local station predicting approaching small craft warnings; for a time, he even played Doc, mugging and ad-libbing his way through his performance. His unprofessional attitude toward his own play and his fellow actors was apparently due, at least in part, to the injections of Ritalin he was taking at the time. His Stoned Age was creeping back.

Williams also participated in a number of post-show discussions. Donald Vining was present at one:

> It was announced that after the play Tennessee Williams himself would answer questions from the stage. His first remark was, "What are you all doing here on a lovely day like this?," which can't give his producers and backers much pleasure. Asked why he had the bartender end by inviting a girl who had repeatedly been described as filthy to stay overnite [*sic*] he said, "Well, I believe anybody is better than nobody. I know some of you don't agree with me, but that's what I believe."[65]

While Williams's existential numbness may have been a personal condition, other gay men in the large urban centers like New York, experienced it, as well. It may not have been politically correct to describe just how anesthetizing sex could become, how like the jabbing of a hypodermic needle it could be, especially in later years looking back, but another writer recorded it, nonetheless:

> Sexual permissiveness became a form of numbness, as rigidly codified as the old morality. Street cruising gave way to half-clothed quickies; recently I overheard someone say, "It's been months since I've had sex in bed." Drugs, once billed as an aid to self-discovery through heightened perception, became a way of injecting lust into anonymous encounters at the baths. At the baths everyone seemed to be lying face down on a cot beside a can of Crisco; fistfucking, as one French savant has pointed out, is our century's only brand-new contribution to the sexual armamentarium. Fantasy costumes (gauze robes, beaded headache bands, mirrored vests) were replaced by the new brutalism: work boots, denim, beards and mustaches. . . . No longer is sex confused with sentiment. . . . Although many gay people in New York may be happily living in other, less rigorous decades, the gay male couple inhabiting the seventies is composed of two men who love each other, share the same friends and interests and fuck each other almost inadvertently once every six months during a particularly stoned, impromptu three-way.

So Edmund White recalled the seventies before they were even over.[66]

───────────

As the 1970s gave way to the eighties and the onset of AIDS, it was not uncommon for some gay men to valorize sex even more than they had done in the wake of gay liberation. It was a way of combating the homo-phobia that accompanied the epidemic. Williams's work, which rarely, if ever, celebrated gay sex as such, would be consigned during these years to the category of self-hatred. Yet, what gay man, even a young gay man, has not feared loneliness, or has not seen an older gay man out cruising, or standing alone in a bar and thought, "I don't want to end up like that" (and for young men, an older man is likely to be anyone who has crossed the threshold of 40)? As Williams's life contracted, loneliness joined his fear of dying alone as the obsession not only of his life but of his work. That it was unfortunate timing for a gay man in the 1970s and 1980s to profess such a fear makes it no less real, for him or for any of his peers. The difference between Quentin and the small, pathetic family that gathers every night in Monk's Place, is that Quentin, alone among them, is aware of his loneliness and his inability to feel surprise; he alone speaks objectively about them. He may not be able to regain what he has lost, and living with his self-knowledge may be more painful than the lives of oblivion the others lead, but the ache of his knowledge offers at least the possibility of redemption.

SIX

"Before My Clean Heart
Has Grown Dirty . . ."

I

Even with the relatively good, or at least not devastating, reception of *Small Craft Warnings,* things got no better for Williams as the 1970s wore on. 1973 saw another Williams play on Broadway, where it most certainly did not belong: *Out Cry,* a revision of *The Two-Character Play,* had been produced in London in 1967 and would be revised several more times, appearing in New York in 1975 as *The Two-Character Play* yet again. *Out Cry* was so personal a play as to be almost inscrutable to anyone not inside Williams's mind. It was certainly beyond the ken of the Broadway audience who sat through it without comprehension for 12 performances. Its two characters, Felice and his sister Clare, are actor-managers who have been deserted by the rest of their company and find themselves locked in a theatre. They perform a play-within-a-play called

The Two-Character Play, which concerns a brother and sister, named Felice and Clare, who are trapped inside the house they have lived in all their lives. *Out Cry* is an endless series of mirrors reflecting Williams's many complex feelings toward his sister, Rose, and an increasing inability to turn the pain of his life into coherent art. Williams thought it was his best play since *Streetcar.* When it closed, Williams fled to Los Angeles to attend a revival of *Streetcar* that starred Faye Dunaway and Jon Voight. In the face of the new work's failure, yet another star-studded revival of the 1948 Pulitzer Prize–winner may have seemed to Williams more a criticism of his current work than appreciation for the old.[1]

Later that month, as a writer who was rich, famous, and constantly surrounded by an entourage, he told a reporter, "I'm a lonely person, lonelier than most people." That loneliness was exacerbated during the first few days of May when his friend the writer Jane Bowles died. A month later Williams was in Rome when he learned that William Inge killed himself. Williams had known Inge since the 1944, when Inge, then the drama critic for the St. Louis *Star-Times,* interviewed him while *The Glass Menagerie* was in rehearsals. They had an intense relationship for a short time that soon cooled into a more casual (and for both men), more emotionally manageable one. Inge gave Williams his early plays to read; in turn, Williams recommended him to Audrey Wood who took him on as a client. Williams watched, first with pride and then with increasing jealousy, as Inge produced hit after hit in the 1950s and early 1960s: *Picnic, Bus Stop, Come Back, Little Sheba,* and *The Dark at the Top of the Stairs.* Unlike Williams, Inge had always been genuinely tormented by his homosexuality; he had rarely known a happy day in his life. Bedeviled by alcohol and drugs, Inge's career slipped away, much as Williams's seemed to. Now that his former friend was no longer a threat to his own fame, Williams felt nothing but sorrow at his death.[2]

From Rome he flew to Tangier, where he gave his sympathies to Paul Bowles; after that it was London and then back to New York. He was there when Anna Magnani, who had starred in the film versions of *The Rose Tattoo* and *Orpheus Descending,* died. Hemmed in by feelings of mortality, he hurried back to Key West. For a decade, Williams had surrounded himself with a revolving door of secretary/companions as well as a coterie of sycophants, mostly young gay men. The secretary/companions seemed genuinely to care for Williams and did their best to insulate him from the chaos he created for himself, but time after time, he would drive them away with impossible demands or personal attacks stemming from his paranoia. In the eyes of Williams's friends, however, the hangers-on were just that: talentless people with little or nothing to recommend them who hung around for drinks, money, and whatever use they could make of the increasingly vulnerable playwright.

David Lobdell witnessed their behavior on several occasions in 1970, when Williams returned from a trip to Asia:

> . . . when Tennessee is here [he wrote his sister Patricia], the parasites seem to flock to his presence, and it is almost like a living nightmare. I'm surprised Tenn puts up with it, but he seems to find it diverting, and when he gets tired of it, he just packs up and takes a long trip. He must know that most of these people are out to take him for all they can get, and that they have nothing to offer him in return, and either he is excessively kind or abnormally foolish to tolerate it."[3]

Despite the increasing loneliness and incessant travel, the writing never stopped. Williams always seemed to have at least half a dozen projects underway at once. Now these included revisions of *The Milk Train Doesn't Stop Here Anymore* for the actress Sylvia Miles and *The Gnädiges Fräulein* for Anne Meacham, who had played Catharine in the first production of *Suddenly Last Summer,* and the new plays, *The Red Devil Battery Sign* and *This Is (An Entertainment),* which seems to have been inspired by Jean Genet's *Splendid's.*[4]

The work that Williams pursued with most energy, however, was his *Memoirs.* As early as 1972, he had completed a draft of 650 pages. A shorter version was published in 1975. If there were any lingering doubts in the public mind about either Williams's sexuality or his openness about it, this book put them to rest. Indeed, to the dismay of many reviewers, the *Memoirs* devotes considerably more space to a candid, if not always accurate, account of Williams's sexual history than it does to his writing. Not everyone shared the disappointment, however. Williams received several hundred letters from gay men ranging from gratitude for

Gentlemen Callers

212

his frankness to requests for dates and meetings. Whatever the book's merits as autobiography, literature, or insight into a life spent in the theatre, *Memoirs* is certainly not the product of a self-loathing mind.[5]

Many thought that Williams in his *Memoirs* was aiming for the same effect that he had tried for recently in his plays: to shock an increasingly unshockable public with material that was not especially shocking. It may be that this attempt was an ill-conceived manifestation of Williams's desire to produce one more talked-about play, to jolt his way back to the top. At about this time, he told an interviewer for the *Saturday Review,* "I'd fuck a baboon to have another hit."[6]

One of the more negative reviews of *Memoirs* appeared in the *Gay Sunshine Journal,* originally a gay tabloid that began publishing in Berkeley, California, in 1970 that, by 1976, had become a literary quarterly. The review's author, Andrew Dvosin, accused Williams's plays of being little more than lies. He repeated the old accusation that Williams's female characters were men in drag. What had once been a staple of homophobic and right-wing critics' and reviewers' attacks now became a weapon in the arsenal of intolerant left-wing gay liberationists.[7]

Williams responded in a letter to the editor that appeared in the next issue, claiming that sexual identity was never a precise thing, and that he'd never met anyone who wasn't androgyne to some degree. He goes on to assert that all the women characters he created were authentically women. Then, however, he blurs the issue by conceding that he and directors sometimes differed on character interpretation, and implying that some of his collaborators chose to emphasize big, universal themes over which gender a specific character may have belonged to subtextually. It seems as if Williams is taking back what he first said, but cloaking it in the ambiguous term of "character interpretation." He reiterates his belief that he has never lied by disguising a male character as a female and then adds that he could say what he needed to say as easily with a straight relationship as with a gay one.

It's a tortured performance, given the simple points he wants to make, reminiscent of his defense of Brick to Walter Kerr as both gay and not gay. Just as in 1955 when he might not have wanted to risk losing the laurels he had won with years of struggle and hard work, it may be that while he honestly believed he had never disguised a male character as a female one, there was a part of him that might have thought he had. Given his old and growing guilt about using others to his own advantage, Williams may have begun believing, after more than a decade of being accused of it in print, that there was some truth to the charge. If he could be guilty of so much betrayal, why not self-betrayal, as well?

Shortly after the review, Williams did an interview for *Gay Sunshine* with the novelist and playwright George Whitmore. He addressed the same issue, again in terms unlikely to win him friends among gay activists, although he considered himself something of one, as well. He never found it necessary, he said, to write a "gay play." "I can get just as much satisfaction, if not more, writing about a love affair between a perfectly normal man and a perfectly normal woman [. . .]" he said. He went on to say that there was never a reason for him to write about a gay relationship—"unless you can interpret that between Skipper and Brick in *Cat* as a love affair, and it's legitimately interpretable that way." Had Williams's thinking on that subject changed or clarified in the years since he'd written Kerr that Brick's sexual adjustment was a heterosexual one, if not one that was "innately 'normal'"? Perhaps Williams had always thought that Brick was gay, and now, for the first time, felt comfortable about saying it publicly. And yet there remain signs of a struggle: he had always gotten satisfaction in writing about "perfectly normal" men and women (and yet, writing in *Memoirs,* he would use quotation marks when referring to heterosexuals as "norms").[8]

At the same time, Williams could be quite direct and practical in responding to the question of why he hadn't written about gay relationships: "There would be no producer for it." Williams reminded Whitmore that until the last decade, "theatre" in America meant the commercial theatre, Broadway theatre, and to succeed in it one had to attract a general audience. "I would be narrowing my audience a great deal. I wish to have a broad audience because the major thrust of my work is not sexual orientation, it's social. I'm not about to limit myself writing about gay people" (313, 314). Other gay playwrights, such as Edward Albee and the emerging Lanford Wilson, felt much the same way, but they weren't doing interviews for gay publications in 1975. Whitmore acknowledges that Williams had written many stories that frankly dealt with homosexual behavior. "I've never had any embarrassment about doing it, never tried to disguise my homosexuality," Williams replied (313). At the age of 65, after having led as open a life as any well-known person of his time, perhaps it struck him as somewhat odd that he was still having to insist that he'd never "faked it."[9]

Yet at the time Williams gave this interview, he was at work on a play that not only had a gay character at its center, but one who was trying to come to grips with his gayness.

II

As Williams insisted on his membership in the gay liberation movement, activists were going from success to success. In 1972, a group of college and university faculty formed the national Gay Academic Union. A year later, a survey by *The New York Times* showed that gay student organizations were being established and officially recognized at 20 colleges and universities in New York, New Jersey, and Connecticut. In July of that year, the New York State Court of Appeals ruled that a person could not be denied admission to the state Bar because he or she was gay.

The most significant milestone, however, occurred on April 8, 1974, when the American Psychiatric Association announced that its membership had voted to remove homosexuality from the list of mental illnesses in its Diagnostic and Statistical Manual of Mental Disorders. The manual had always categorized homosexuality as a disease and a form of sexual deviance. Now, no longer considered by psychiatrists as a disease that could—or should—be treated, homosexuality was replaced with a new term, "sexual orientation disturbance," which described the condition of those whose attraction to their own gender caused them mental distress. When the APA Board of Trustees had affirmed the decision the previous December, they were at pains to point out that the new category was "distinguished from homosexuality, which by itself does not necessarily constitute a psychic disorder."[10]

The Board's decision to ratify the recommendation of its Nomenclature Committee was not universally popular within the APA. Opponents of the change, who believed that both the committee and the Board had caved into pressure from activist groups, obtained the required 200 signatures to force a referendum by the association's membership. The announcement of April 8 said that 5,854 members voted in favor, while 3,810 were opposed. The total was about half of the APA's entire membership. The dispute about the demedicalization of homosexuality would go on for years, inside and out of the APA, but the symbolic importance of the event was enormous.

The formation of the Gay Academic Union, the new National Gay and Lesbian Task Force, the gay student organizations, and the APA's decision were all results of the new attitude of activism that characterized at least some lesbians and gay men, most of whom lived in cities. The gays and lesbians who began the student groups were inspired by the gay liberation movement and, in some cases, were responding to specific displays of homophobia on their campuses or in nearby towns. Barbara Giddings and Frank Kameny, two pioneering activists who had organized the first gay rights marches, in Washington and Philadelphia, had for years been appearing at APA conventions to make the case that lesbians and gays were not, by definition, sick. While a number of activist-led breakthroughs would eventually fizzle due to internecine disputes between moderates and radicals, a new era for gays and lesbians was being established. Even if specific initiatives and organizations foundered, many more lesbians and gays came to believe that they could change their lives through political action. The decision to come out itself became a political act.

The theatre was not exempt from these changes, and they would not occur without controversy and rancor. The question was asked: in a time of consciousness-raising, an emerging community and a coalescing of new identities, what was the role of theatre? In practical terms, what part should gay playwrights play? Should they employ their talents for "the good of the community" (which would depend on who defined "community" and "good") by constructing role models and "positive images"? Or should artists go their own ways, creating art in whatever form with whatever meaning in answer to their own private imperatives? And should playwrights be concerned about the "message" or images that their characters might send about gays and lesbians to the world at large?

The answer was yes to both positions—and also no. In 1973, a rancorous debate broke out in the pages of the *Times* over a new musical by

Al Carmines called *The Faggot*. It opened in June at the Truck and Ware-house Theatre, just a year after *Small Craft Warnings* debuted there. With *The Faggot*, Carmines had created a revue that celebrated gay life or trivi-alized it, depending on which side of the argument one favored. One sketch featured a young man who has joined a group called The Dis-senters, which has pledged to oppose any popular viewpoint. He has also joined the "Passive Caucus of the Gay Activists Alliance," and, whatever the situation, prefers to feel desperate. "Some people just *like* being des-perate," he says, adding that being a desperate gay man is a full-time job. Another skit showed five bored and jaded middle-aged gay men sitting in a bar complaining about life when the New Boy in Town struts in. Each of the older men propositions him in ways guaranteed to appear degrad-ing; each is summarily rejected. A number of sketches had to do with ca-sual sex in movie-houses and street corners. There were also skits featuring two mothers *kvelling* over their sons' impending nuptials to each other; Gertrude Stein and Alice B. Toklas sang a homey song cele-brating "Ordinary Things," and Carmines himself played Oscar Wilde in a turn about Wilde and Bosie Douglas.

When he wrote *The Faggot,* Carmines said that he didn't ask himself whether it would "help the gay cause," be "politically liberated and cor-rect," or whether it would "adequately explain the gay life to 'straights.'" "I am not concerned," he wrote in the *Times,* "with the 'image' gay peo-ple or straight people would like to project. If I wanted to deal in those images, I would be in advertising, not theatre. . . . My domain—if I have one—is that crack between ideologies where contradictory, frustrating, unideological, stinking and thrilling humanity raises its head."[11]

Martin Duberman, scholar, playwright, activist, and co-founder of the Gay Academic Union, was unamused. He thought *The Faggot* trivial-ized the pain of gay life. In an article in the Sunday Arts & Leisure sec-tion of the *Times,* he called the play "an affront." True, he admitted, "We all need a laugh. And if it happened to be at our own expense, if it con-firms the social stereotypes that have made our lives as gay people a laugh a minute, well then, here's to those other light-hearted entertainers: Stepin Fetchit, Charlie Chan and the Bloody Injun."[12]

Jonathan Katz, whose play, *Coming Out!,* was championed by Duber-man as one that, as opposed to *The Faggot,* demanded an end to oppres-

sion, wrote to the *Times* that Carmines' play added to the oppression of
gays by reinforcing stereotypes and failing to account for any of the social
pressures "which have led to homosexual self-hatred and desperation."

> There exists at this particular time in our history what seems to me a mar-
> velously inspiring and challenging role for the homosexual artist: to create
> a new, liberated gay culture which is both of high artistic quality and re-
> flective of the new consciousness being created by gay liberationists.[13]

Katz did not note that *Coming Out!*, an admittedly agit-prop play written
to arouse its gay audience to united action for change, also had been attacked by some activists as not being "positive" enough.

The poet Alfred Corn, in another letter to the *Times,* suggested

> Not only must a theatrical work about gay life in 1973 necessarily have a
> political dimension, it must even have political consequences because of
> its effect on a large audience. The dimension [that *The Faggot* has] is the
> negative one of omission. If you don't challenge the present view of homo-
> sexuals, you reinforce it.[14]

Where did Tennessee Williams stand in this argument? He said nothing publicly; his actions suggest he was predictably ambivalent. At the
same time that he viewed himself as "a founding father of the gay movement," he insisted that he could not make political statements in a direct
way. The first position was more wish than fact: He did not participate in
public protests or debate. However, coming out on television, it might be
argued, was a form of activism and suggested that Williams had moved
from the position suggested by *Cat on a Hot Tin Roof.* Now, perhaps, he
was willing to risk his early-won laurels by doing what no other major
American playwright had done: state publicly that he was gay.

That he could not make political declarations directly in his work
says, as categorically as Williams could say it, that he was an artist first.
For an artist, he suggests, the synthesis that Katz asked for was rarely
possible. Williams responded to his own inner urges, not to external
agendas.

As far as gay critics were concerned, it seemed that Williams was so
out of date or irrelevant that he was not worth mentioning in the context
of "the new consciousness being created by gay liberationists." Michael

Smith's *Village Voice* review of *Small Craft Warnings* was a single, sad paragraph that didn't mention the play's gay characters (perhaps he felt he was being charitable not to do so) or indeed anything that suggested that the play had anything to do with or say about life in 1972.

III

The battles fought in Williams's last plays are with acceptance, loneliness, and guilt. David Lobdell reported to his sister that, after all these years, Williams was still afraid of being alone at night. Given Williams's life in the early and middle 1970s, it is not surprising that the primary circumstance of *Vieux Carré* is loneliness.[15]

Vieux Carré takes place in a dilapidated boarding house in New Orleans in 1938. The action is seen through the eyes of a narrator, a young man called simply The Writer, Williams's vision of himself in 1938. In a limited sense, the play is about The Writer's coming of age, his acceptance of his true sexual nature, leaving a lonely home and launching himself into the world. All the characters in *Vieux Carré* are lonely people near the end of their ropes. They are exhausted, the young characters as well as the old, worn out by a world that has proved to be too much for them.

"I'll tell you," says the ancient landlady Mrs. Wire, "there's so much loneliness in this house that you can hear it. Set still and you can hear it: a sort of awful—soft—groaning in all the walls." Loneliness is the illness from which all the characters suffer, from which they are all trying to flee. "Why, there's a saying," the tubercular painter Nightingale tells The Writer in Scene Two, "'Better to live with your worst enemy than to live alone.'" When Tye, a fag-bashing stud like Bill in *Small Craft Warnings,* objects to his girlfriend Jane spending time with a queer like The Writer, she stands on her rights. "I will not have this young grifter who has established squatter's rights here telling me that I can't enjoy a little society in a place where—frankly I am frantic with loneliness!" Mrs. Wire, in Scene Seven, explains her drinking to The Writer:[16]

> I only touch this bottle, which also belonged to the late Mr. Wire before he descended to hell between two crooked lawyers, I touch it only when

forced to by such a shocking experience as I had tonight, the discovery
that I was completely alone in the world, a solitary ole woman cared for
by no one. You know, I heard some doctor say on the radio that people
die of loneliness, specially at my age. They do. Die of it, it kills 'em. Oh,
that's not the cause that's put on the death warrant, but that's the *true*
cause. (65)

Mrs. Wire can hear loneliness groaning in the walls, even if The
Writer cannot. However, she says, that is only because he is too young
to hear it. He will. In one moment of action without words, the
young Writer expresses his loneliness as a young Williams might have:
He caresses the warm sheets of his bed where a young man has lately
laid.

The loneliness The Writer feels is not a result of his homosexuality, of
which he is newly aware, but is part of the callousness he senses is already
beginning to deprive his heart of feeling. "Oh—there's a price for things,
that's something I've learned in the Vieux Carré. For everything that you
purchase in this market-place you pay out of *here!* [*He thumps his chest.*]
And the cash which is the stuff you use in your work can be overdrawn,
depleted like a reservoir going dry in a long season of draught . . ."
(43–4). Like Quentin in *Small Craft Warnings*, The Writer is already
being gnawed at by the loss of the ability to be surprised. One might ask
if in the world of the play such doubt is earned. Whether it is or not,
Williams remained obsessed with this image of himself in 1977, when
Vieux Carré was produced on Broadway.

The play bears a striking resemblance to *The Glass Menagerie*. This is
not so surprising, given the success of several revivals of his early plays at
the time Williams was at work on this new one. It is told to us within a
story frame provided by a narrator who is a young writer yearning for
freedom whose initials are TW. Its characters have the same overall objective, or spine, as those in *The Glass Menagerie:* to defeat a hostile world,
to keep at bay outside elements threatening destruction. In *Vieux Carré*
the principal means they employ to achieve this is denial. The painter
Nightingale insists that he suffers from a cold when in fact he is dying of
tuberculosis; Mrs. Wire, the landlady, must believe that she runs the only
respectable boarding house in the French Quarter; Jane, a young fashion
illustrator from the North, refuses to face the blood disease that is slowly

killing her. Only The Writer, who at first denies his homosexuality, comes to some sort of acceptance.

Mrs. Wire is an older, more desperate version of Amanda. Her husband deserted her long ago; now she must make ends meet by herself. She can be shrill and controlling, but also remarkably understanding. In Scene Six she appears dressed just as Amanda does in Scene Three of *The Glass Menagerie*. The stage direction for the latter reads, "*Now we see Amanda, her hair is in metal curlers and she is wearing a very old bathrobe, much too large for her slight figure, a relic of the faithless Mr. Wingfield*" (162). In Scene Seven of *Vieux Carré*, Mrs. Wire describes her own appearance in the previous scene: "[. . .] my hair in curlers, me wearin' the late, long ago Mr. Wire's old ragged bathrobe" (64).

The relationship between The Writer and Mrs. Wire echoes the one between Tom and Amanda. Mrs. Wire is nosy and puritanical; she absolutely must know what her boarders are up to at all times and is outraged by any hint of sexual impropriety—meaning sex of any sort. She enlists The Writer in her latest scheme by hiring the destitute young man to distribute advertisements ("Meals for a Quarter in the Quarter") for her little lunchroom. As demanding as Amanda, eventually she insists he spend more time on the job than he is inclined to do. The Writer reacts as Tom would—by quitting.

Like Tom, The Writer yearns to break free, but for most of the play does not know how. He devotes himself to his writing. When escape comes, it is in the form of a young man. A musician named Sky invites The Writer to ride west with him in his beat-up 1932 Ford (just as Williams left New Orleans with the musician named Jim Parrott). As Amanda pleads with Tom not to leave until Laura is married and taken care of, so Mrs. Wire begs The Writer to stay:

> You won't [leave] if I can prevent it and I know how. [. . .] I'm gonna inform your folks of the vicious ways and companions you been slipping into. [. . .] Address and phone number, I'll write, I'll phone!—You're not leavin' here with a piece of trash like *that* that pissed out the window!— Son, son, don't do it! (77)

Although she can't admit it, she needs him to stave off loneliness. The Writer replies, "Mrs. Wire, I didn't escape from one mother to look for another" (77). Four scenes later, The Writer tells Mrs. Wire, "I'm not your child. I am nobody's child. Was, maybe, but not now. I've grown into a man about to take his first step out of this waiting station into the world" (107). The moment of flight, however, holds more terror for him than it did for Tom Wingfield. The difference is the effect of Williams's old adversary, time. Tom's escape is a liberation. The Writer, created 33 years after *The Glass Menagerie*, is much more wary:

> Mrs. Wire: Now watch out, boy. Be careful of the future. It's a long ways for the young. Some makes it and others git lost.
> Writer: I know . . . [He turns to the audience.] I stood by the door uncertainly for a moment or two. I must have been frightened of it . . .
> Mrs. Wire: Can you see the door?
> Writer: Yes—but to open it is a desperate undertaking . . . ! (116)

The urge for flight, always a part of Williams's make-up and a crucial trait in so many of his characters, was so strong that it withstood the years of bitter experience. In the end, The Writer goes. Before he does, however, he stands before the open door and hears "the waiting storm of his future—mechanical racking cries of pain and pleasure, snatches of song" (116). The older Williams sees his younger self staring into the future and, skipping over the years of happiness, of surprise, of love with Frank Merlo, and tremendous professional accomplishment, imagines being transported directly to a Stoned Age, reaping the punishment of experience without the rewards of the experience itself. *The Glass Menagerie* was, in part, a look backward. *Vieux Carré* is a look forward to a time when the past is remembered only from a vantage point of sadness and regret. By the mid-1970s, Williams had come a long distance from Tom Wingfield, and, over the intervening 30 years, had walked into many bars to find companionship.

The first encounter between The Writer—young, reticent, sexually inhibited—and Nightingale—experienced, older, sexually frank, and voracious, desperate for human contact—recalls in a curious way the Gentleman Caller scene in *The Glass Menagerie*, but here the roles are reversed. In *The Glass Menagerie,* the Gentleman Caller was interested in neither love nor sex, while Laura was head over heels in love. Nightingale, who comes calling on The Writer, is looking for both love and sex, while The Writer unsuccessfully tries to put him off. During the day, Nightingale does pastel portraits of tourists at a place called The City of Two Parrots. He is a male Hannah Jelkes. Will he bring peace to The Writer?

The scene is gentle and tender, offering sex as an intimate palliative for loneliness. The scene is tinged with guilt, as befits both The Writer and Williams in New Orleans in 1938, but it ends on notes of love and forgiveness and a Chekhovian refusal to judge by conventional moral standards.

The scene begins with The Writer's sobs of loneliness. Nightingale, responding to the cries he hears through the thin plywood partitions that separate their rooms, offers advice:

> Nightingale: Trying not to, but crying . . . why try not to? Think it's unmanly? Crying is a release for man or woman . . .
> Writer: I was taught not to cry because it's . . . humiliating . . .
> Nightingale: You're a victim of conventional teaching, which you'd better forget. (18)

If the writer is shy and inhibited, the painter is not. "A single man needs visitors at night," Nightingale says. "Necessary as bread, as blood in the body" (20). In this first scene of gay sex that Williams had produced, in his first play in which a character's gay sexuality is at the center of the action, sex between men is presented not as something furtive or shameful; it is offered as a sacrament. If The Writer does not accept it initially as such, this is not another indication of homophobic discourse or "heterosexism," but an honest recollection of the way things were and another installment in the ongoing debate between Williams and himself over acceptance of his sexuality. The debate is not an indication of self-hatred alone, but one more example of a lifelong attempt to overcome it. There is nothing but tenderness and longing in The Writer's recollection to Nightingale of his homosexual initiation with a paratrooper who seduced him on New Year's Eve beneath a sunlamp.

Nightingale, for all his loneliness and denial of his terminal condition, is also a man of dignity and wisdom. As The Writer mourns the loss of his grandmother, Grand, Nightingale gives him a comforting pat: "Well, losses must be accepted and survived" (19). When Nightingale suggests that The Writer have an operation to remove a cataract even though he cannot pay for it and The Writer responds that he could not be so dishonest, the older painter's response, "Don't be so honest in a dishonest world," is not an invitation to step into the closet (19). It is the advice of a man who has lived in a world in which people are constantly using others. It is significant, too. that this gentle artist who wishes to teach The Writer something about himself should be named for a bird. He harkens back to the canaries transformed from goldfish by a rainbow-colored scarf, and to the canaries that the Baron de Charlus longs to hear in his mattress. Nightingale represents the escape from the stultifying and the conventional that homosexuality often represents in Williams's plays.

In *Acting Gay,* Clum is hard on *Vieux Carré* and on this scene. In it, he writes:

> The narrator, like Tom in *The Glass Menagerie,* ruefully remembering events from his young manhood, focuses on speculations regarding his landlady's judgment of his sexual activity with the artist, Nightingale: "I wonder if she'd witnessed the encounter between the painter and me and what her attitude was toward such—perversions? Of longing?" It is interesting that Williams's autobiographical narrator seems to internalize the judgmental woman's attitude toward homosexual acts, as it is interesting that he has to imagine a judgmental audience for his sexual initiation. (165)

This judgment would be something to reckon with were it an accurate report of what happens in the scene. But the witness is not Mrs. Wire, it is the ghost of Grand, The Writer's grandmother, and she is far from judgmental. After The Writer and Nightingale have had sex (which is not his "sexual initiation"), The Writer has a vision:

> Now it was [my grandmother] who stood next to my bed for a while. And as I drifted toward sleep, I wondered if she'd witnessed the encounter between the painter and me and what her attitude was toward such—perversions? Of longing? [. . .] Nothing about her gave me any sign. The

weightless hands clasping each other so loosely, the cool and believing gray eyes in the faint pearly face were as immobile as statuary. I felt that she neither blamed nor approved the encounter. No. Wait. She . . . seemed to lift one hand very, very slightly before my eyes closed with sleep. An almost invisible gesture of . . . forgiveness? . . . through understanding? . . . before she dissolved into sleep . . . (26–7)

The sexual encounter between Nightingale and The Writer is gentle and based on both a loneliness and a desire for comfort that is mutual.

The sex between men and women, however, as shown and described in *Vieux Carré,* is a brutal expression of power. Another Gentleman Caller is expected in Scene Nine, but not in the benevolent guise of Nightingale. He is a Brazilian businessman whom Jane met at the Blue Lantern Bar. He mistakes her for a prostitute, and given her desperate financial state, she decides not to disabuse him. She invites him back to the boarding house. When Jane demands that Tye, her petty thief and drug addict boyfriend, move out, he rapes her.

In Scene Eleven, Tye tells Jane of the murder of a stripper at the strip joint where he works as a barker. The Champagne Girl, as she was called, had told her boss that she was leaving for the West Coast. This gentleman, called The Man, had a different idea. Rather than let her escape (she is another one desperate to break away), he has her devoured by dogs. Those who believe that exotic behaviors such as cannibalism are a special punishment reserved for predatory homosexuals such as Sebastian Venable need to account for this heterosexual perversion of longing. More important, the Man would rather kill the Champagne Girl than let her go. The murder is a cry, *in extremis,* of loneliness. As for Tye, when he awakens from a drunken stupor in The Writer's cubicle to find both Nightingale and The Writer hovering over him, we learn exactly how heterosexual he is, and how flexible he is in his attitude toward homosexuals: "No goddamn faggot messes with me, never! For less'n a hundred dollars!" (45). The critics who have obsessed over Williams's supposed homophobia rarely seem to consider the "negative images" of fag-bashers that Williams included in his last plays.

The time separating Tom Wingfield of *The Glass Menagerie* from The Writer of *Vieux Carré* is more than just the few years it took Tennessee Williams to get from St. Louis to New Orleans. The two young men are separated by the 33 years of their author's tumultuous, difficult life. *Vieux Carré* is a young man's beginnings reflected not by a young man but by an older, sadder, experienced man, the product of a culture that is homophobic, puritanical, materialistic, and anti-art, not to mention the result of a repressed, highly neurotic early family life and a recent, almost fatal battle with depression. Unlike Tom Wingfield, who seems to write prolifically, The Writer needs jump-starting, and already fears the irreplaceable emptying out of his personal material, his emotional resources, that will lead to a life in which surprise, amazement, and creation are no longer possible.

▬▬▬▬▬

Williams had written about homosexuality all of his working life. Sometimes, references were inferential, as in *Auto-Da-Fé* and *The Glass Menagerie;* other times, he was quite explicit, as in *A Streetcar Named Desire, Camino Real,* and *Suddenly Last Summer.* Almost always, these plays were shaped by the pressure of his twin needs to conceal and reveal. This condition, which produced most of Williams's best work, vanished, more or less, as he emerged from his Stoned Age. Some have said his "new" willingness to write openly gay characters and place them at or very near the center of his plays was a result of the social movements of the 1960s, and in part this was doubtlessly so. His newfound relative lack of reticence (for Williams could never be considered a reticent writer), however, extended to areas other than sexuality. Indeed, what made, for many audiences, his plays of the late 1960s and onward so difficult to watch was not only their formal difficulties, but the excruciating amount of existential pain Williams made no attempt to conceal. Beginning in 1966, with *Slapstick Tragedy,* the various incarnations of *The Two-Character Play* (1967) and *In the Bar of a Tokyo Hotel* (1969) are plays that are practically unwatchable due to the sheer level of torment Williams inflicts on his characters, reflecting the pain he felt before drug-induced numbness set in. The tight artistic control over difficult material that he had demonstrated with *Suddenly Last Summer* and to only a slightly lesser degree in *The Night of the Iguana* had dissolved, along with the elements of his life that had kept him stable. The breakdown of his relationship with Frank

Merlo and then Merlo's death brought Williams close to the precipice of madness, and the drugs and alcohol he pumped into his system in the years to come loosened his control over his work even more. Another way of looking at what happened to Williams's work in the 1960s is to see these plays as desperate searches for a craft and a form to contain what seemed to him a holocaust of pain. In *The Gnädiges Fräulein* (part two of *Slapstick Tragedy*), Williams mixed spiritual brutality with farce. In *In the Bar of a Tokyo Hotel* he reined in his habitual loquaciousness and used the fewest words to contain the greatest psychic distress. The same man who wrote what some considered to be the melodramatic excesses of *Battle of Angels, Orpheus Descending*, and *Suddenly Last Summer* created the stark, monochromatic, caged-in world of the dying artist and his wife in a country far from home and almost beyond understanding. In place of the long, florid speeches heavy with imagery, or the narrative monologue, are sentence fragments. Speech, it seems, cannot express or contain the artist's fear and pain. Neither, any longer, can conventional dramatic action: there is virtually none in *In The Bar of a Tokyo Hotel*. (It resembles, as much as anything, Eugene O'Neill's *Welded*, in which a tormented married couple play out their love-hate relationship in two spotlights, just as Williams's characters exist inside circles of lights surrounded by darkness.) Mark's and Miriam's physical and emotional world has been reduced literally to recriminations and cruelties. Mark's pain is brought on by what appears to be the disappearance of his talent: for artists like him and Williams, there is no real division between artist and man. When one dies, what, necessarily, must become of the other?

What some have seen as Williams's response to Stonewall, gay liberation, and other movements was more a reaction to the inner movements of his personal and artistic needs. Williams was correct when he told Arthur Bell that he had never faked it, had neither disguised gay characters as heterosexual ones nor ever pretended to be anything he wasn't. Now, several conditions were leading him to reveal himself, not more than before, but in different ways. The forthright speech of Quentin is one way—but how more forthright is this than the Baron's actions in *Camino Real* in 1953? Williams's several candid interviews during the 1970s were another. His last plays with gay characters, including *Vieux Carré* and *Something Cloudy, Something Clear,* deal as openly about them

as any latter-day critic or activist could demand, but in the end, they are
reflections of Williams's inner explorations and attempts at achieving
self-forgiveness, not of an outside political or aesthetic agenda. Williams
was criticized, and rightly, for often placing too much importance on the
critical reception his plays received (like his fellow Missourian Mark
Twain, Williams wanted approval at least as much as he wanted to rebel);
still, he followed his own dictates. As an artist, even a faltering one, he
could do no less.

While The Writer is as reticent as Tom when it comes to describing
his nocturnal habits, the play makes very clear the general subject that is
on his mind: sex. This, paradoxically, accounts for one of the reasons
that *Vieux Carré*, while a stronger play than its critics contend, is less
powerful than *The Glass Menagerie*. Little of a sexual nature is hidden in
the later play; but *The Glass Menagerie* is dimly lit. Post-Stonewall critics
who wish that Williams had been more politically correct and had created more "affirmative images" of gay men fail to realize that much of
Williams's power as a writer derived from a strong impulse not to reveal.
When society decreed that perhaps a little bit of open homosexual life
might be shown on mainstream stages and Williams complied, an important source and condition of his strength as an artist evaporated.
What Clum and de Jongh regard as Williams's personal failing is nothing less than an indication of Williams's artistic victory. If it is true that
much art has as its source the early pain of its creator, then Williams put
his pain to great artistic use. He turned the negative impulses of the
closet into art. The motor that drives most of his best work is the conflict between a never-quenched desire to reveal sexual truth and all the
social and personal conditions that militated against such revelations.
Psychological security, even in relative terms, is not always a fertile condition for the creation of art. In the 1970s the cultural demand for silence had been partially lifted, and there was suddenly less for Williams
the artist to struggle against. Habit remained, of course. Williams was
58 when the Stonewall Riots occurred (and that event changed little
overnight); he had a lifetime of cultural hatred to struggle against and
struggle he did.His deep ambivalence about relationships and companions is not an indication of hatred of his homosexuality. Looking at the
totality of Williams's life, it is much more likely that the ambivalence is

about allowing anyone, of any gender or sexuality, to see into his heart. To fall in love was to risk giving up something he was loathe to lose. It is expressed in his picture of gay men and gay relationships because he was gay. If Williams had been straight, he would have exhibited it in his portrayals of straight relations. In fact, it exists there constantly: Val and Lady, Stella and Stanley, Shannon and any number of women and girls, Kilroy and the Gypsy's Daughter. All of these relationships are animated by ambivalence, often to the point of violence and even death.

IV

By 1981, Williams was struggling on several fronts. *Vieux Carré* had closed after seven performances; his next play on Broadway, in 1980, *Clothes for a Summer Hotel*, failed, as well. Williams still had not absorbed the lesson that there might be better venues for him than Broadway. In between these two productions came a production at Florida Keys Community College of a play that he'd been working on since at least 1969, *Will Mr. Merriweather Return from Memphis?* Despite its moments of poetry, the play confused more people than it pleased. Quickly, it was returned to the worktable. In 1979, *A Lovely Sunday for Crève Coeur* opened at the Off-Broadway Hudson Guild Theatre where it eked out 36 performances. The same year, Eve Adamson, the artistic director of the Jean Cocteau Repertory Company, a small Off Off-Broadway non-profit theatre that produced mostly classic plays on the Bowery in New York City, became interested in Williams's new work. She produced what he subtitled "An Outrage for the Stage," the grotesque and bawdy (but not especially funny) *Kirche, Kutchen und Kinder*, as a work-in-progress, to keep the critics away. The Kennedy Center Honor he received at the end of the year, and the Medal of Freedom he received at a White House Ceremony the following year (one week after the death, at 95, of the fearsome Edwina), did little to assuage his growing sense of failure. Over the years, his paranoia only seemed to increase, as well, so that eventually it was almost impossible to keep secretaries or assistants for any length of time. His unwarranted suspicions, along with the demands of keeping his increasingly chaotic life in order, chased them away in rapid succession. The paranoia may have been in part hereditary, but it was not helped by the reappearance of drugs. By the early 1980s, the Nembutal and Seconal had returned to his medicine bag, along with Ritalin.[17]

Into the 1980s Williams continued to grapple with the problem of finding a dramatic form suitable to express the existential condition of his life while critics wondered why he bothered. *Something Cloudy, Something Clear* is, like *Vieux Carré,* and again *The Glass Menagerie,* a memory play. Unlike its predecessors, however, which, despite their story frames that occur in the present are set largely in the past, *Something Cloudy, Something Clear* dispenses with conventional stage notions of time. The action occurs in different time periods simultaneously.

Two characters, August, a young playwright, and Clare, a young woman, exist in 1940 and 1980 simultaneously; both experience their shared past and observe it from the future. Often, Williams does not specify where in time they are. In a sense, they are everywhere at once. "Life is all—it's just one time," says the leading character. "It finally seems to all occur at one time."[18]

If little is veiled in *Vieux Carré,* nothing at all is hidden in *Something Cloudy, Something Clear.* The action is almost entirely about the relationships between August, Clare (another heterosexual dying of a then-fatal disease, diabetes), and the sexually ambiguous dancer Kip, who is also dying, of a slow-growing brain tumor. Clare appears to be an entirely fictional character; August is a self-portrait and Kip a rendering of Kip Kiernan.

Once again, the prevailing condition is loneliness. The action consists of the things one does to combat loneliness that inevitably injure others; the authorial judgment is forgiveness. What the play lacks in strong dramatic forward movement—difficult to achieve in a play told in non-linear time—it makes up for in a broad vision of life. The setting is ostensibly the beach near Provincetown, Massachusetts, but is really the windswept landscape of Williams's imagination in which characters from various periods of his life appear, almost in the free-associative way of a memory wandering across the events of a lifetime.

The action begins almost immediately. August is typing in his beach shack when Clare and Kip appear over the dunes. Kip is a Canadian draft dodger, and Clare a young woman on the lam from a notorious gangster named Bugsy Brodski. They recognize August as the man who appeared before them drunk the night before and stared so long at Kip that the dancer was afraid he would be seduced. Still, Kip senses immediately that August might have his uses. Summer is ending, the artists for whom Kip poses for money will be dispersing; Clare's "keeper," Brodsky, will be coming to claim her. Neither Clare nor Kip want her to return to this thuggish gangster, but they know that they both will soon need to be cared for. "Perhaps some other—some alternative to it exists," Kip muses. "We could be two kept for the price of one" (3). Can they use August without hurting him too much?

For his part, August has already fallen in love with Kip. But he is dubious that either loving or being loved are propositions in which he can believe for very long, and wonders to what extent he merely wants to use Kip to satisfy his insatiable sexual appetite. Indeed, as August stares hard at Kip, the dying Frank Merlo appears in a wheelchair, and the implicit question arises, to what extent did Williams think he used him? Merlo was his lover and also his secretary and the person who made daily existence possible for the time they were together. Kip Kiernan occupied a month in Williams's life of seven decades; Merlo was the anchor of Williams's existence during 14 years of his greatest creative activity. They are linked here through Williams's feelings of guilt.

August himself is on the verge of making a concession for the sake of survival, an artistic compromise that he hopes will guarantee the production of his play by the pompous Broadway producer Maurice Fiddler and his wife Celeste (stand-ins for the Theatre Guild's Lawrence Langner and his wife, actress Armina Marshall). The name tells us all we need to know of what Williams thought of Langner and his middlebrow pretensions to Art. They appear on the dunes to negotiate terms for a production of August's new play (read, Williams's *Battle of Angels*). Maurice speaks in vampiric terms of the need for the new theatrical blood that August represents. He wants to pay the young playwright a $50-a-month option, while August demands the standard Dramatists Guild minimum of $100. Negotiations ensue, just as they did between Langner and Audrey Wood. Fiddler and August both make much of standing on principle, but what is at issue is money: Fiddler has it; August wants it.

Something Cloudy, Something Clear is neither romantic nor sentimental. It portrays life as a continual jockeying for advantageous position. There is something predatory in all of the characters' negotiations (including those between August and a belligerent sailor over sex), but Williams finally is able to understand and forgive the fact that negotiation for advantage is an unavoidable part of life if one is to live in the world. Indeed, most of the action of the play involves bargaining, the giving of one thing in order to gain another. Even the intoxicated sailor, who earlier had stumbled away from August's shack feeling cheated because August refused to be fucked, comes stumbling back at the end of Part One willing to make a deal: "[. . .] you can fuck me for another fin and a drink" (56). This is a proposition August can easily agree to.

The first scene in Part Two is a protracted negotiation between August and Kip. Kip, knowing the "exigencies of desperation" (62), is bound to give in; both he and August sense this and August presses his advantage. In the end, Kip will surrender to August's "amatory demands," making it clear he'd rather not, but knowing he must. It is not a pretty scene, and Williams does not spare August nor allow him to look especially chivalrous. But what, Williams also asks, are the alternatives? Sex in *Something Cloudy, Something Clear* is neither joyous nor freeing. It is little more than a bargaining chip in a negotiation, a commodity to be bought and sold. Kip, who gives in to August's desires only reluctantly, regards sex (*all* sex, not just gay sex) as something that animals do: He sounds rather like Amanda in *The Glass Menagerie*.

Bugsy Brodksy is the only character in the play unwilling to bargain; he merely takes what he wants without negotiation or payment. If this is the only alternative to compromise and bargains in the real world, Williams suggests, then compromise must be accepted as a human, even a charitable, response to unfavorable circumstance, even if it comes with a price. Brodsky is another in Williams's late line of fag-bashers. As the play's one representative of unrepentant heterosexuality, he is no recommendation for that particular "lifestyle." He is brutal, exploitative, and not above using violence to get his way. He—and the predatory kind of sex he stands for—is one of the dangers from which the characters seek shelter.

Along with understanding comes a hard-won forgiveness. If August feels guilt for not always behaving admirably, he suggests that using people for one's own advantage is, for better or worse, natural—as natural as having sex. There is little anger in the play, but there is regret: regret that the world is not better than it is.

From the vantage point of 1980, August sees more clearly than he did in 1940 (when his left eye was clouding over with a cataract) that compromise is a necessary component of life if one is not to live alone and lonely. This may be neither a pleasant nor an "affirmative" truth, but it is the way things are. August's most precious possessions are a silver Victrola (like Laura Wingfield's) and a battered typewriter (like Tom's) and all his negotiations and compromises are undertaken in order to cling to these symbols of his goodness and purity. What would he do, he asks Clare, if he were to lose them? Battling the Fiddlers into accepting his revisions and paying him the full $100 a month, he finds that he can fight and survive. All that separates August from Kip and Clare in terms of morality is early death. They died too young to be corrupted by compromise, just as Leona's brother Haley did in *Small Craft Warnings*. Haley died before he could use others; Kip and Clare before they had to make more ugly compromises with the world. August lives on: sullied, impure, but surviving. He sees in 1980 that for him there was no other choice but to get down into the muck of the "I'll give you this for that" (32). Even the young August seems to see it, and regrets it ruefully as an older man would. He recites a poem he says he wrote at 16 that begins,

God, give me death before thirty,
Before my clean heart has grown dirty [. . .]
Soiled with the dust of much living,
More wanting and taking than giving . . . (23)

August receives visits from two other great loves of his life. The first is a young woman named Hazel (based on Williams's childhood friend Hazel Kramer), whom August loved only platonically. When he tells her that her appearance in the play comes too late for him to love her physically, she tells him a surprising secret: She has always loved girls. August tells her that the fact that she simply *loved* is sufficient, while she in turn helps him see that the guilt he felt over drilling a hole in a swimming pool cubicle to watch the boys shower was almost too ludicrous to require forgiveness.

The second love to appear is Frank Merlo. In *Something Cloudy, Something Clear,* what Williams still seems unable to forgive himself for is his "desertion" of Merlo while he was dying. It was a charge Williams angrily denied when it appeared in print in the 1970 *Atlantic* interview; for months he obsessed over it in letters to Maria St. Just. Yet it surfaces here again, years later. Of all the moments Williams might have chosen from his 14 years with Merlo, the one he depicts is the one that seems guaranteed to inflict the most pain and guilt on his conscience: when Frank, dying of cancer, "turned away from me on the Memorial Hospital bed and pretended to sleep [. . .]" (17).

━━━━━━━

If Williams had harbored only negative feelings toward homosexuality he probably would not have written plays. Instead, he may well have tried to kill himself, marry, or do what so many other gay men did, especially in the 1950s: seek a psychoanalytic "cure." That he had such feelings cannot be denied—it would have been almost impossible for a gay man or lesbian born in 1911 and raised in the deep South and Middle America not to have had them—but one can see, from *The Glass Menagerie* through *Something Cloudy, Something Clear,* a lifelong struggle against them. Slowly, he brought homosexuality from a place hidden and offstage—but a vital and palpable element nonetheless—to the very visible center of his late plays. He waged his battle in the only place a playwright of his generation could: on the commercial stage, where in the middle decades of the twentieth century all the odds were against him. Yet at the height of the McCarthy and HUAC witch-hunts he put on Broadway an openly gay man in *Camino Real;* a few years later he outlined the moral paralysis of a man afraid to admit his homosexuality for fear of losing all the fame and love he had gained; a few years after that, he described the fate of a man who dared to live at the edges of sexual adventure and paid the price. In the 1970s and 1980s he used gay characters to describe surviving the almost unendurable personal pain that burned him deeply, even as he emerged from it largely untransformed.

When Frank Merlo appears in *Something Cloudy,* he tells of a compromise. When asked to confess, Frank agrees, to please the priest, although he himself cares little for religion. While the Church regards Frank's and August's relationship as a cardinal sin, Frank tells the priest that his "sin" is as white as his blood is red. The fact that love between men is "forgiven" as a small, white sin might be considered proof enough that Williams was, to the end, brimming with self-hatred. Why, an affirmative image critic might ask, must same-sex love be regarded as a sin of any size or color? Williams might respond that that was how homosexuality was viewed by the world he knew, and drama can take place nowhere but in the world. That world, the complex, real one, rarely deals in matters that can truthfully be reduced to, or expressed in, simple "affirmative" or "negative" images. Drama that is not merely melodrama admits them even less often. These kinds of images are the province of propaganda. Propaganda might be art in the proper circumstances, but it was never Williams's kind of art. Honest, humane judgments are not easily arrived at in a world where the grays predominate. This is the world in which Tennessee Williams lived and about which he wrote, even as that world slowly began to change. He wrote about the world in terms he knew. These terms were neither homophobic nor heterosexist. They were complex. His guilt went deeper than and beyond his homosexuality, which he wrote about in terms that challenged, not conformed to, the conventional thinking of the era of his growth and development. Williams's guilt was rooted in part in his feelings for his puritanical mother and promiscuous father and the deep and complicated love that he felt for his sister. If Williams had any distaste for sex, it was for sex generally, not homosexual acts. His guilt stemmed also from what he perceived as his own callousness and from the society in which he was raised.

The last image of *Something Cloudy* is hopeful and valedictory. Just before Clare and Kip fade away, August scribbles a note. "What was that you just wrote?" Clare asks. August responds, "Just a note of reference for tomorrow" (85). The imagining, the re-fashioning of chaotic experience into something that suggests meaning, will go on. "How did it go," August asks himself, "that bit of Rilke? 'The inscrutable Sphinx? Posing forever—the human equation—against the age and magnitude of a universe of—stars . . . '" (85). There is no answer to that equation; it cannot be found in the reductive formulation of the "positive" or "negative" image or even in art. " . . . and while this memory lives, the lovely ones remain here, undisfigured, uncorrupted by the years that have removed me from their summer" (85). No matter how stained he felt himself to be by using the pain of others for his own artistic ends the writing would go on, in the hope that through the creation of order and meaning might be found some comfort, even salvation. Williams waged his battle long before Stonewall, long before gay academic unions, before the advent of "gay pride" as an imaginable notion. He cannot be condemned for writing about his world and experience instead of one he was not born into and about emotional experience he did not have. To do so would be to give in to an outlook far narrower than Williams's and a judgment far harsher than any he ever accorded his gay characters.[19]

Notes

INTRODUCTION

1. Harold Clurman, *All People Are Famous* (New York: Harcourt Brace Jo-
   vanovich, 1974), 230.
2. Walter Kerr, *Journey to the Center of the Theater* (New York: Alfred A.
   Knopf, 1979), 181.

CHAPTER 1

1. Lyle Leverich, *Tom: The Unknown Tennessee Williams* (New York: Crown,
   1995), 420.
2. Ibid., 285; Don Lee Keith, "New Tennessee Williams Rises from 'Stoned
   Age,'" in *Conversations with Tennessee Williams,* ed. by Albert J. Devlin
   (Jackson and London: University Press of Mississippi, 1986), 151.
3. Tennessee Williams, *Vieux Carré,* in *The Theatre of Tennessee Williams* Vol.
   8 (New York: New Directions), 50 (except for *Cat on a Hot Tin Roof,* all
   quotations from Williams's plays are taken from these editions, only the
   number of which will be noted hereafter. Subsequent references to plays
   previously cited will be noted in the text).
4. Tennessee Williams, *Memoirs* (New York: Anchor Press/Doubleday, 1983), 12.
5. Leverich, *Tom,* 107.
6. Ibid., 308; Donald Windham, ed., *Tennessee Williams' Letters to Donald
   Windham, 1940–1965* (Athens and London: The University of Georgia
   Press, 1996), 9–10.

7. Donald Windham, "Tennessee Williams: Humpty Dumpty Before, During, and After the Fall," *Christopher Street,* Issue 94, 1985, 50–1.
8. Leverich, *Tom,* 373.
9. Williams, *Memoirs,* 25, 29–30, 45; Leverich, *Tom, fn,* 604–5; St. Just, *Five O'Clock,* 235.
10. St. Just, *Five O'Clock,* 336; Williams, *Memoirs,* 241.
11. New Orleans *Times-Picayune,* September 2, 1941,1; September 4, 12; November 7, 47; Leverich, *Tom,* 422.
12. Leverich, *Tom,* 425.
13. Leverich, *Tom,* 421.
14. Ibid., 428.
15. Ibid., 429.
16. Ibid., 430.
17. *Theatre of TW,* Vol. 8, 155.
18. Tennessee Williams, "Mornings on Bourbon Street," in *The Collected Poems of Tennessee Williams,* ed. by David Roessel and Nicholas Moschovakis (New York: New Directions, 2002), 72.
19. W. Kenneth Holditch, "The Last Frontier of Bohemia: Tennessee Williams in New Orleans," *Southern Quarterly* 23:2 (Winter 1985), 12.
20. Leverich, *Tom,* 301.
21. Ibid., 301.
22. Tennessee Williams, *Auto-Da-Fé,* in *Theatre of TW,* Vol. 6, 131.
23. Tennessee Williams, *The Glass Menagerie* in *The Theatre of TW* Vol. 1, 164, 173.
24. Allan Bérubé, *Coming Out Under Fire: The History of Gay Men and Women in World War Two* (New York: Plume, 1991), 15.
25. Ibid., 19–20.
26. Ibid., 20–21.
27. Donald Vining, *A Gay Diary, 1933–1946* (New York: Hard Candy Books, 1996), 283.
28. Eric Bentley, *The Brecht Memoir* (Evanston: Northwestern University Press, 1989), 112–113; Eric Bentley to Michael Paller, interview, June 6, 1997.
29. Bérubé, *Coming Out,* 21, 25.
30. Windham, *Letters,* x; Leverich, *Tom,* 449.
31. Vining, *Gay Diary, 1933–1946,* 321, 326, 457.
32. Leverich, *Tom,* 477.
33. Ibid.
34. Williams, *Memoirs,* 99.
35. Leverich, *Tom,* 477.
36. Tennessee Williams, *Orpheus Descending.* In *The Theatre of Tennessee Williams* Vol. 3 (New York: New Directions, 1971), 271, 305.

37. Williams, Tennessee. Untitled. Tennessee Williams Papers, Rare Book and Manuscript Library, Columbia University.
38. Leverich, *Tom*, 491.
39. Williams, *Memoirs*, 78.
40. Louis Kronenberger, "A Triumph for Miss Taylor," *New York News PM*, April 2, 1945; Burton Rascoe, "'The Glass Menagerie' An Unforgettable Play," *New York World-Telegram*, April 2, 1945; Otis L. Guernsey, Jr., "The Theatre at Its Best," *New York Herald Tribune*, April 2, 1945; John Chapman, "'Glass Menagerie' is Enchanting Play, Truly Hypnotic Theatre," *New York Daily News*, April 2, 1945, all qtd. in *New York Theatre Critics' Reviews*, 1945, pp. 234–7.
41. Roger Boxill, *Tennessee Williams* (New York: St. Martin's Press, 1988), 69.
42. Ibid., 35–6.
43. Ruby Cohen, "The Garrulous Grotesques of Tennessee Williams," in *Modern Critical Views: Tennessee Williams*, ed. by Harold Bloom (New York: Chelsea House Publishers, 1987), 59.
44. Leverich, *Tom*, p. 174; Robert Rice, "A Man Named Tennessee," *New York Post*, May 4 1958, M–2.
45. Tennessee Williams, "The Catastrophe of Success," in *The Theatre of TW* Vol. 1 (New York: New Directions, 1971), 140.
46. Tennessee Williams, *The Glass Menagerie*, in *The Theatre of TW* Vol. 1 (New York: New Directions, 1971), 143. Further references are in the text.
47. Mark Lilly, "*The Glass Menagerie* and *A Streetcar Named Desire*," in *Lesbian and Gay Writing*, ed. by Mark Lilly (London: Macmillan, 1990), 153–63.
48. Williams, *Memoirs*, 84.
49. Igor Stravinsky, *Poetics of Music in the Form of Six Lessons*, trans. Arthur Knodel and Ingolf Dahl (Cambridge: Harvard University Press, 1975), 63.

CHAPTER 2

1. James H. Jones, "Dr. Yes," *The New Yorker*, August 25th/September 1, 1997, 99–113.
2. Ibid.; John D'Emilio, *Sexual Politics, Sexual Communities: The Making of a Homosexual Minority in the United States, 1940–1970* (Chicago: University of Chicago Press, 1983), 37; Jones, 107.
3. D'Emilio, *Sexual Politics*, 36.
4. Ibid., 37.
5. Will Roscoe, ed., *Radically Gay: Gay Liberation in the Words of its Founder Harry Hay* (Boston: Beacon Press, 1996), 60–1.
6. Ibid.
7. Ibid., 62.
8. Ibid., 64–5.

9. Ibid., 82–3.
10. Ibid., 61.
11. Jones, "Dr. Yes," 100.
12. Bérubé, 262; D'Emilio, *Sexual Politics,* 44–5; Vining, *A Gay Diary* Vol. 2, 288.
13. D'Emilio, *Sexual Politics,* 41.
14. Ibid., 44; Jonathan Katz, *Gay American History: Lesbians and Gay Men in the U.S.A.* (New York: Thomas Y. Crowell Co., 1976), 93.
15. D'Emilio, *Sexual Politics,* 46–7.
16. Ibid., 50.
17. Martin Bauml Duberman, "Hunting Sex Perverts," *Christopher Street,* 5, no. 12 (1982): 46.
18. Ibid., 46–7.
19. Ibid., 48, 47.
20. Ibid., 48.
21. Ibid.
22. Katz, *Gay American History,* 100–2; D'Emilio, *Sexual Politics,* 44.
23. Qtd. in Nicholas von Hoffman, *Citizen Cohn: The Life and Times of Roy Cohn* (New York: Doubleday, 1988), 229.
24. Since 1927, it had been illegal to produce plays in New York State "depicting or dealing with the subject of sexual degeneracy, or sex perversion." The penalties included the padlocking of the theatre where the play was produced. See Curtin, 100, for details. The law was not revoked until 1968.
25. Qtd. in Kaier Curtin, *We Can Always Call Them Bulgarians* (Boston: Alyson Publications, 1987), 293–4.
26. Mordaunt Shairp, *The Green Bay Tree,* in *Gay Plays, Volume One,* ed. Michael Wilcox (London: Metheun, 1985), 55.
27. Qtd. in Curtin, 187.
28. Ibid., 188.
29. John Chapman, "'The Green Bay Tree' Is Revived," *Daily News,* February 2, 1951; Richard Watts, Jr., "Decline of 'The Green Bay Tree,'" *New York Post,* February 2, 1951, qtd. in *New York Theatre Critics' Reviews,* 1951, 368–70.
30. Shairp, 85.
31. Bérubé, 257–9.
32. Roscoe, 135–6.
33. Gene Heil, "The Darker Side of the Fifties," *Christopher Street,* vol. 2 no. 3, September 1977, 14–6.
34. Ibid.
35. Lillian Hellman, *The Children's Hour,* in *Twenty Best Plays of the American Theatre,* ed. John Gassner (New York: Crown Publishers, 1965), 595.

36. Thomas R. Dash, "The Children's Hour," *Women's Wear Daily,* December 19, 1952.
37. Bérubé, 269–70.
38. Robert Anderson, *Tea and Sympathy* (New York: New Dramatists, 1983), 86, 88. Subsequent references are in the text.
39. William Hawkins, "Pilloried Innocence Sets Dramatic Plot," *New York World-Telegram,* October 1, 1953; "'Tea and Sympathy' a Beautiful Play; Deborah Kerr Magnificent," *Daily News,* October 1, 1953; Richard Watts, Jr., "The Arrival of a New Playwright," *New York Post,* October 1, 1953, qtd. in *New York Theatre Critics' Reviews,* 1953, 151–3.
40. Doris Kearns Goodwin, *No Ordinary Time: Franklin and Eleanor Roosevelt: The Home Front in World War II* (New York: Simon and Schuster, 1994), 556.
41. Tennessee Williams, Foreword, *The Theatre of Tennessee Williams* Vol. 2 (New York: New Directions, 1971), 419–20.
42. Tennessee Williams, *Camino Real, The Theatre of Tennessee Williams* vol. 2 (New York: New Directions, 1971), 435. Further quotations are in the text.
43. William Hawkins, "'Camino Real' Is Pure Emotion," *Journal-American,* March 20, 1953; Clurman, *Divine Pastime,* 21–2.
44. Georges-Michel Sarotte, *Like a Brother, Like a Lover: Male Homosexuality in the American Novel and Theatre from Herman Melville to James Baldwin* (New York: Anchor Press/Doubleday, 1978), 110–11.
45   Ibid., 111.
46. David Savran, *Communists, Cowboys and Queers: The Politics of Masculinity in the Work of Arthur Miller and Tennessee Williams* (Minneapolis: University of Minnesota Press, 1992), 81–2.
47. Ibid., 91.
48. Robert J. Corber, *Homosexuality in Cold War America: Resistance and the Crisis of Masculinity* (Durham: Duke University Press, 1997), 116; Clum, *Acting Gay* (Durham: Duke University Press, 1997), 162, 166.
49. Brenda Murphy, *Tennessee Williams and Elia Kazan* (Cambridge and New York: Cambridge University Press, 1992), 67.
50. Williams, *Memoirs,* 98.
51. Qtd. in Devlin, 31; Williams, Foreword to *Camino Real, Theatre of TW* Vol. 2, 419.
52. Spoto, 208.
53. Qtd. in Devlin, 219; Williams, *Orpheus Descending, Theatre of TW* Vol. 3, 305.
54. Ibid., 292.
55. Williams, *Five O'Clock,* 336.
56. Harold Clurman, *The Fervent Years* (New York: Da Capo, 1983), 151.

57. Elia Kazan, *A Life* (New York: Anchor Books, 1989), 494–98; Williams, *Five O'Clock*, 56.
58. Simon Karlinsky, ed. *Anton Chekhov's Life and Thought,* Michael Henry Heim, Simon Karlinsky, trans. (Berkeley: Univ. of California Press, 1973), 117.

CHAPTER 3

1. John D'Emilio, *Making Trouble: Essays on Gay History, Politics and the University* (New York and London: Routledge, 1992), 63.
2. Williams to Paul Bigelow, April 5, 1950. Tennessee Williams Papers, Rare Book and Manuscript Library, Columbia University.
3. Williams to Paul Bigelow, April 12, 1950. Tennessee Williams Papers, Rare Book and Manuscript Library, Columbia University.
4. Spoto, 210–12.
5. D'Emilio, *Sexual Politics,* 50.
6. Ibid.
7. Windham, 290.
8. Qtd. in Curtin, 310.
9. Ruth and Augustus Goetz, *The Immoralist* (New York: Dramatists Play Service, 1980), 44, 45. Subsequent references are noted in the text.
10. Eric Bentley, *What is Theatre?* (New York: Limelight Editions, 1984), 150.
11. Ibid., 151.
12. John Chapman, "'The Immoralist' Sure Enough Is," *Daily News,* February 9, 1954. *New York Theatre Critics Reviews,* 1954.
13. Tennessee Williams, "Three Players of a Summer Game," in *Collected Stories* (New York: New Directions, 1985), 305–6. Subsequent references are in the text.
14. Tennessee Williams, *Cat on a Hot Tin Roof* (New York: Signet Books, 1955), 45. Subsequent references to this edition are in the text.
15. Windham, 16.
16. Murphy, 104.
17. Ibid., 107.
18. Ibid., 105.
19. Robert Coleman, "'Cat on a Hot Tin Roof' is Likely to be Hit," *Daily Mirror;* John Chapman, "'Cat on a Hot Tin Roof' Beautifully Acted, but a Frustrating Drama," *Daily News;* Richard Watts, Jr., "The Impact of Tennessee Williams," *New York Post;* John McClain, "Drama Socks and Shocks," *Journal-American;* William Hawkins, "Cat Yowls High on 'Hot Tin Roof," *New York World-Telegram,* all March 25, 1955, qtd. in *New York Theatre Critics' Reviews,* 1955, 343–4.
20. Walter F. Kerr, "Cat on a Hot Tin Roof," *Herald Tribune,* March 25, 1955.

21. Williams, *Five O'Clock Angel*, 109.
22. Ibid., 110.
23. Ibid., 110–12.
24. Tennessee Williams, *Cat on a Hot Tin Roof* (New York: Dramatists Play Service, 1958), 28.
25. Nicholas de Jongh, *Not in Front of the Audience* (London and New York: Routledge, 1992), 71.
26. Qtd. in Devlin, 40.
27. Williams, *Memoirs*, 167.
28. D'Emilio, *Sexual Politics*, 51.
29. Qtd. in Kenneth Lewes, *The Psychoanalytic Theory of Male Homosexuality* (New York: Simon and Schuster, 1988), 136–37; Jones, 110.
30. Spoto, 227.
31. Clum, *Acting Gay*, 156. See also de Jongh, Savran and Corber. For example, to Savran, Big Daddy's youthful indiscretions with hobos and plantation owners are now costing him dearly. His fatal colon cancer is, in Savran's interpretation of Williams's outlook, "the wages of sodomy." This illness is, "as in 'The Mysteries of the Joy Rio,' . . . the currency of mortal debt in Williams's homosexual economy." While Big Daddy also has a streak of libertarianism in him, Williams (in this view) has planted even deeper "a malignancy that is unmistakably the sign of his homosexual inheritance." (Savran 101) Unmistakable it may be, perhaps, to Savran and other academic critics who comb literature for examples of colon cancer suffered by homosexual characters and apply them as an all-embracing trope to any writer of any era regardless of personal experience and sensibility. Such a symbol is not, unmistakably, a trope of Tennessee Williams's. Indeed, other than Big Daddy and the two characters of "The Mysteries of the Joy Rio," Emil Kroger and Pablo Gonzales (whose bowel cancers can also be interpreted not as the wages of sodomy but those of hiding from the world one's true nature), there is not another gay character in Williams's plays that succumbs to it. The artist Nightingale will die of tuberculosis in *Vieux Carré* and Haley, the offstage gay brother of Leona in *Small Craft Warnings*, dies of pernicious anemia. Williams's other gay characters, unless one counts the bisexual Kip in *Something Cloudy, Something Clear*, do not suffer a metaphorical punishment from a wrathful God or from nature. (Lest one think, however, that only gay characters die of painful diseases in Williams's plays, numerous heterosexual characters succumb, as well, including Lord Mulligan in *Camino Real*, Lot Ravenstock in *Kingdom of Earth*, Trinket in *The Mutilated*, Mark in *In the Bar of a Tokyo Hotel*, D. H. Lawrence in *I Rise in Flame, Cried the Phoenix*, Mrs. Shapiro in *This is the Peaceable Kingdom*, Clare in *Something Cloudy, Something Clear*, and Jane in *Vieux Carré*. Among the gay characters who do not die

of bowel cancer in Williams's plays: Sebastian Venable in *Suddenly Last Summer,* Tom Wingfield, The Writer in *Vieux Carré,* August in *Something Cloudy, Something Clear,* Quentin in *Small Craft Warnings,* and the Baron de Charlus.)

32. Williams, *Memoirs,* 171; *Five O'Clock,* 131.

CHAPTER 4

1. Windham, 292.
2. Qtd. in Devlin, 52.
3. Williams, *Memoirs,* 172.
4. Nancy M. Tischler, *Tennessee Williams: Rebellious Puritan* (New York: Citadel Press, 1961), 246; qtd. in Devlin, 51; Williams, *Memoirs,* 169; Tischler, 247; Spoto, 239.
5. Tennessee Williams, "The Man in the Overstuffed Chair," in *Tennessee Williams: Collected Stories* (New York: New Directions, 1985), xv-xvi.
6. Qtd. in Lewes, 31–2.
7. Ibid.
8. Qtd. in Richard A. Isay, *Being Homosexual: Gay Men and Their Development* (New York: Avon, 1989), 6.
9. Bérubé, 229–57.
10. Isay, 6–7.
11. Eugene B. Brody, interview with author, November 1, 1999; Isay, 7. Not all members of this establishment were blind to or content with this transformation of a once radical movement. In 1953, Clarence P. Oberndorf, who 30 years earlier had been president of the American Psychoanalytic Association, wrote, "Psychoanalysis had finally become legitimate and respectable, perhaps paying the price in becoming sluggish and smug, hence attractive to an increasing number of minds which found security in conformity" (Isay, *Being,* 136).
12. Richard A. Isay, *Becoming Gay: The Journey to Self-Acceptance* (New York: Henry Holt, 1996), 20.
13. Qtd. in Lewes, 128; ibid., 129.
14. Edmund Bergler, *Homosexuality: Disease or Way of Life?* (New York: Hill and Wang, 1957), 188; and *One Thousand Homosexuals: Conspiracy of Silence, or Curing and Deglamorizing Homosexuals?* (Paterson: Pageant, 1959), 7; Bergler, *One Thousand,* 9; Bergler, *Homosexuality* 33–48; Bergler, *Homosexuality,* 9 and *One Thousand,* viii; qtd. in Lewes, 113; Ibid., 114; Bergler, *Homosexuality,* 200; qtd. in Lewes, 131; Delos Smith, "Critics Warn of Harm in Kinsey's 'Fallacies,'" *New York Post,* January 7, 1954, 36; Bergler, *One Thousand,* 5.
15. Both qtd. in Lewes, 131.

16. Bergler, *Homosexuality,* 28–9.
17. Ibid., 175.
18. Qtd. in Lewes, 165.
19. Bergler, *One Thousand,* 249.
20. Bergler, *Homosexuality,* 78–9.
21. John Crowley, letter to author, October 1, 1999.
22. Eugene B. Brody, "Symbol and Neurosis: Selected Papers of Lawrence S. Kubie," *Psychological Issues* 11, no. 4 (1978): Monograph 44, 2.
23. Brody interview; Glover, 12–13; Brody, "Symbol," 5.
24. Brody, "Symbol," 6; Ibid., 1; Glover, 8.
25. Lawrence S. Kubie, *Practical and Theoretical Aspects of Psychoanalysis* (New York: International Universities Press, 1950), xiii.
26. Lawrence S. Kubie, *Neurotic Distortion of the Creative Process* (New York: Noonday Press, 1961), 63. There is no knowing, of course, if the playwright Kubie refers to is Williams. It may be him; it may be Inge. It may not be a playwright at all, if Kubie, in the interest of keeping his clients' sessions confidential, created a playwright where, say, a novelist existed.
27. Ibid., 64.
28. Kubie, *Practical,* 88–9.
29. Lawrence S. Kubie, Letter to Tennessee Williams. January 13, 1958. Rare Book and Manuscript Library, Columbia University.
30. Spoto, 240 and Williams, *Memoirs,* 173; Kubie, *Practical,* 86–7.
31. Kubie, *Distortion,* 4.
32. Kubie, *Distortion,* 24 and 39. In *Touched with Fire: Manic-Depressive Illness and the Artistic Temperament* (New York: Free Press, 1993), a study of artists and mental illness, Kay Redfield Jamison (who categorizes Tennessee Williams as a manic depressive) discusses the similarities between mildly manic moods and creativity: "Two aspects of thinking in particular are pronounced in both creative and hypomanic thought: fluency, rapidity, and flexibility of thought on the one hand, and the ability to combine ideas or categories of thought in order to form new and original connections on the other. The importance of rapid, fluid and divergent thought in the creative process has been described by most psychologists and writers who have studied human imagination" (105).
33. Kubie, *Distortion,* 39, 60.
34. Brody, interview; Arthur Laurents, *Original Story By* (New York: Alfred A. Knopf, 2000), 72 and John M. Clum, *Something for the Boys: Musical Theatre and Gay Culture* (New York: St. Martin's Press, 1999), 103.
35. Brody, *Symbol,* 9.
36. Tennessee Williams, *Suddenly Last Summer, The Theatre of Tennessee Williams* Vol. 3 (New York: New Directions, 1971), 351. Subsequent references are in the text.

37. Edwina Williams, *Remember Me to Tom* (New York: G. P. Putnam's Sons, 1963), 244.
38. Qtd. in Devlin, 50.
39. Ibid., 51–2, 169.
40. Edwina Williams, 242–3.
41. Leo H. Bartemeier, "In Appreciation." *The Journal of Nervous and Mental Disease* 149 no. 1 (1969): 19.
42. Kubie, *Distortion,*10; qtd. in Devlin, 40.
43. Qtd. in Leverich, 174; Brody, "Symbol," 26; qtd. in Edwina Williams, 137.
44. Dating this play is (as is so often the case with Williams) vexatious. The typescript, written on stationery of the Comodoro Hotel and the Robert Clay Hotel in Miami, is undated. On June 13th, Williams wrote to Sandy Campbell on Comodoro stationery. (Windham, 293). On a note appended to the manuscript in the UCLA archives, he wrote, "Written in Havana shortly before Castro regime." Castro seized power on January 1, 1959.
45. Tennessee Williams, *And Tell Sad Stories of the Deaths of Queens* in *Political Stages: Plays That Shaped a Century,* Emily Mann and David Roessel, eds. (New York: Applause, 2002), 394. Subsequent quotations are noted in the text. Although the editors write that the play takes place either between 1939 and 1941, or between 1945 and 1947, it is clear from the text that it is set at the time when Williams wrote it, most likely 1957. References to big televisions and hi-fis suggest that the editors' dates are too early, as does the chronology of Candy's relationship with Sidney: It began 18 years previous to the play "in the war years . . ." (395). If the play takes place in 1957, then 18 years earlier would be 1939—not quite the war years for America, but certainly for Europe.
46. Leverich, 477.
47. There is a tantalizing connection between *And Tell Sad Stories of the Deaths of Queens . . .* and *Suddenly Last Summer.* In the Humanities Research Center at the University of Texas at Austin, is a file labeled *Meeting of People.* It contains a short scene (also written on the stationery of an unidentified Havana hotel) between a middle-aged woman and a young sailor she picked up the previous night in the famous French Quarter gay bar, Lafitte in Exile. In the untitled scene, the sailor, named Casky, accuses the woman of stealing his money, although it seems clear that he had none to begin with. He is hung over, discomfited, and nervous while the woman is amused and self-possessed. The scene is undated, but may have been a start to what became *And Tell Sad Stories . . .* , featuring a woman as the sailor's seducer rather than a gay man. Here is a possible example of Williams turning a straight female character into a gay male one—just the

reverse of what Williams was often accused. Of course since the manuscript, like that of *And Tell Sad Stories* is undated, the reverse may also be
true. If that's the case, one might speculate that Williams, after finishing
And Tell Sad Stories, began it again with the gender of the seducer changed
to fit the conventional mores of the 1950s—and then changed his mind,
satisfied with the first version. Or, the untitled scene might be a complete
one-act in itself, or even meant to go with *And Tell Sad Stories,* as a variation of that latter relationship. Williams's chaotic work habits make speculation seductively endless. The characters' names are interesting, too.
"Casky" will become "Buck" and then "Karl" in *And Tell Sad Tales,* while
the woman's name and gender change from Mrs. Venable to Candy.
"Candy" acquires some of "Casky's" toughness in the transformation.
Also, "Mrs. Venable" suggests that the fragment was written around the
same time as *And Tell Sad Stories,* before *Suddenly Last Summer.* Conversation with Nicholas Moschovakis, June 2, 2004, who provided me with the
text of the untitled scene.

48. See also *The Last of My Solid Gold Watches.* Written sometime prior to
1946, it is a sympathetic portrait of a traveling salesman very much like his
own father.

49. Williams, "Overstuffed Chair," x, vii.

50. Edwina Williams, *Remember Me,* 26.

51. Tennessee Williams, "Something Unspoken," in *The Theatre of Tennessee
Williams* Vol. 6 (New York: New Directions, 1981), 282. Subsequent references are in the text.

52. Tennessee Williams, "Williams's Wells of Violence," *New York Times,*
March 8, 1959, II 1, 3.

53. "Of the Deep South and Yemen and Ionesco," *Cue;* "Garden District,"
Theatre Arts, March 1958, 13; Henry Hewes, "Theatre," *Saturday Review,*
January 25, 1958; Brooks Atkinson, "Theatre: 2 by Williams," *New York
Times,* January 8, 1958, 23.

54. Kenneth Tynan, "In Darkest Tennessee," *Observer,* September 21, 1958.

55. Qtd. in Edwina Williams, *Remember Me,* 48.

56. de Jongh, 82; John Clum, "'Something Cloudy, Something Clear': Homophobic Discourse in Tennessee Williams" in *Displacing Homophobia: Gay
Male Perspectives in Literature and Culture,* ed. Ronald R. Butters, John M.
Clum, and Michael Moon (Durham and London: Duke University Press,
1989), 157.

57. Tischler 261–2. Neither David Savran nor Robert Corber, in their attempts to portray Williams as a revolutionary fit for late-twentieth-century
political correctness, can bring themselves to mention the play at all.

58. Tennessee Williams, "Suddenly Last Summer," in *The Theatre of Tennessee
Williams* Vol. 3 (New York: New Directions, 1971), 375, further references

are in the text; and Clum, "Something Cloudy," 157–8. See Windham, 215.

59. Qtd. in Edwina Williams, 137.

60. As late as 1977, Williams complained, in a joint interview with William Burroughs, about the literalness of the film version: "When you began to see Mrs. Venable, and it became so realistic, with the boys chasing up the hill—I thought it was a travesty. It was about how people devour each other in an *allegorical* sense" (qtd. in Devlin 304).

61. Ibid., 423.

62. In an undated draft, the set description in typescript reads, "Very delicate and free. It should not be clear whether the acting area is in or outside the house, it may be a glass-walled 'conservatory' or it may be a patio or parts of both. It is flanked by a garden which is like a formalized jungle with intermittent bird cries." This fragment is headed with the alternative title, *Cabeza de Lobo* or *A Las Cinco de la Tarde* (Butler Library, Columbia University).

63. In another undated, unnumbered fragment, different from the one cited above, in the Williams collection at the Butler Library Rare Manuscript and Book Collection at Columbia University, the ending is unequivocal, and the doctor refuses to operate: "As for the girl's story, well, it's a sick girl's story, it's wildly visionary . . ."

> (Jungle music fades in, very softly)
> —*Not* literally accurate, I hope. But—the images of it had a meaning, a truth, that a literally accurate account of what happened to your Cousin Sebastian in Cabeza de Lobo last summer probably wouldn't have had [. . .]—It *was* a hideous story! But I'm not going to try to cut it out of her head, not for all the subsidies in the world!
> (He picks up his little black bag as—
> THE CURTAIN FALLS SLOWLY
> The End.)

64. Lawrence Kubie letter to Tennessee Williams, January 13, 1958. Tennessee Williams Papers, Rare Book and Manuscript Library, Columbia University. All subsequent quotes from Kubie to Williams are in this letter.

65. Arthur Gelb, "Williams Explains His Move Off Broadway," *New York Times*, December 16, 1958; "Williams, "Web of Violence"; "Of the Deep South"; Elliot Norton, "Tennessee Williams Goes Ghastly in New Play," *Boston Daily Record*, February 26, 1958.

66. "Of the Deep South"; Brooks Atkinson, "'Garden District'," *New York Times*, January 19, 1958, II, 1 ; George Oppenheimer, "The Higher Depths," *Newsday*, January 17, 1958, 7; "Two by Two," *Time*, January 20, 1958, 42; "Garden District," *Theatre Arts*, March 1958, 13.

67. Edwina Williams, 13.

CHAPTER 5

1. William Hoffman, *Gay Plays: The First Collection* (New York: Avon, 1979), Introduction, xxiii–xxiv.
2. One play that had its first production at the Playwrights Unit was Mart Crowley's *The Boys in the Band.* (Gussow, 187)
3. Hoffman, xxviii.
4. Ibid., ix–x.
5. Steven Samuels, "Charles Ludlam, a Brief Life," in *The Complete Plays of Charles Ludlam* (New York: Perennial Library, 1989), xii.
6. Michael Smith, "The Good Scene: Off Off-Broadway," *The Drama Review* 10, no. 4 (1966), 176. For a comprehensive history of Off-Off Broadway see Stephen J. Bottoms, *Playing Underground* (Ann Arbor: Univ. of Michigan Press, 2004).
7. Albert Poland and Bruce Mailman, eds., *The Off Off Broadway Book* (New York: Bobbs-Merrill, 1972), xxv.
8. When Charles Ludlam described his own theatre as a synthesis of "wit, parody, vaudeville farce, melodrama, and satire," he could have been describing Williams's efforts from the late 1960s onwards, as well, including *Slapstick Tragedy, This Is* and *Kirche, Kutchen and Kinder.* That theatre, however, did not exist for Williams in the 1940s and 1950s, and by the time he was writing work that resembled Ludlam's description in tone and subject matter, it was by then an over-familiar genre for which Williams, no matter how great his sympathies for it, lacked both the requisite light touch and the refusal to be serious that formed cornerstones of this theatre's aesthetic. Indeed, Williams's traditional kind of earnestness was frequently the target of these plays' satire. The style that Williams would evolve beginning in the middle 1960s, while more than familiar to downtown audiences, was radically strange to the upper- and middle-class audience who, in decreasing numbers, attended his plays on Broadway (Samuels, ix).
9. Lewis Funke, "News of the Rialto," *New York Times,* November 5, 1961, II, 1.
10. Qtd. in Devlin, 164–5,101.
11. Ibid.,104.
12. Spoto, 272–3.
13. Tennessee Williams to Frank Merlo, Tennessee Williams Papers, Rare Book and Manuscript Library, Columbia University.
14. Spoto, 269.
15. Richard Gilman, *Common and Uncommon Masks: Writings on Theatre, 1961–1970* (New York: Random House, 1971), 144.
16. Tennessee Williams to Frederick Nicklaus, August 14, 1963, Tennessee Williams Papers, Rare Book and Manuscript Library, Columbia University.

17. Williams, *Five O'Clock,* 185.
18. Ted Kalem, "The Angel of the Odd," *Time,* March 9, 1962, 54.
19. For an excellent analysis and history of *Tokyo Hotel,* see Allean Hale, "*In the Bar of a Tokyo Hotel:* Breaking the Code," in *Magical Muse: Millennial Essays on Tennessee Williams,* ed. Ralph F. Voss (Tuscaloosa: Univ. of Alabama Press, 2002), 147–62. Also, Michael Paller, "The Day on Which a Woman Dies: *The Milk Train Doesn't Stop Here Anymore* and Noh Theatre," in *The Undiscovered Country: The Later Plays of Tennessee Williams,* ed. Philip C. Kolin (New York: Peter Lang, 2002), 25–39. Stefan Kanfer, "White Dwarf's Tragic Fadeout," *Life,* June 13, 1969, 10. *New York Times,* June 10, 1969, 96.
20. Tennessee Williams, *Now and at the Hour of Our Death.* MS. dated 1970 in the Special Collections archive of UCLA, 3, 8, 14. Subsequent quotations are noted in the text. Another, very similar draft is stored in the Rare Books and Manuscript Department of Butler Library at Columbia University, dated 1970.
21. David Savran sees the fragmentary language of *In the Bar of a Tokyo Hotel* as an attempt by Williams to find a "new" language of "obscurity or indirection" to hide his homosexuality since his former theatrical language could no longer do so in the face of plays such as *Boys in the Band.* Of course, Williams had not hid anything about his homosexuality since *Camino Real* of 1953. At the same time, Savran also sees the language as a rebellion against the war in Viet Nam and other social injustices being fought out in the1960s. Savran, 136–7.
22. Williams was uncertain about the play's possibilities. He wrote in his stage directions, "*I have not yet decided whether or not 'The Hunched Man in Black' bearing a placard [. . .] is a legitimate stage device and I doubt that I'll know till I see the play performed. [. . .] Most of the dialogue between MADGE and BEA should be paced like a ping-pong ball. There's probably too much of it but they intrigue me as characters so I find it hard to cut them down to their probable right proportion in the play*" (2). Despite these doubts, Williams gave the play to Audrey Wood, who had an agency title page put on the chaotic draft, which includes several crossed-out and rearranged passages. Nothing seems to have happened with it until 1980, when, according to Allean Hale, Williams's then agent, Mitch Douglas, submitted a shortened version, along with two other one-acts (*A Cavalier for Milady* and *The Youthfully Departed*) to New Directions for inclusion in the seventh volume of *The Theatre of Tennessee Williams.* The latter play was set in type and then withdrawn, while *Cavalier* didn't make it that far into the production process. The new version of *Now and at the Time of Our Death* was called *Now the Cats with Jewelled Claws.* The men's dialogue was substantially rewritten (and their names removed, replaced with "First Young

Man" and "Second Young Man") and another character, The Manager, ele-
vated from minor status to sinister supporting role—a simpering, unattractive gay man who proves to be the surviving First Young Man's future.
There is also an interesting clue provided as to the identity of the members
of the Mystic Rose. The First Young Man says, "Remember, the first Mystic Rose ended up high, too, on crossed beams between two thieves." (Tennessee Williams, *Now The Cat With Jewelled Claws,* in *The Theatre of
Tennessee Williams* Vol. 7 [New York: New Directions, 1981], 324)
Williams asked New Directions to withdraw *Now the Cat,* saying he'd been
ill when he'd written it, but the play was already in print. Email from Allean Hale to author, May 23, 2004.

23. Donald Webster Cory and John P. LeRoy, *The Homosexual and His Society:
A View From Within* (New York: The Citadel Press, 1963), 3. By 1973,
Sagarin had changed his mind and, writing under his real name, affirmed
that a cure was indeed possible and desirable. See Martin Duberman, "The
'Father' of the Homophile Movement" in Duberman, *Left Out: The Politics
of Exclusion/Essays/1964–1999* (New York: Basic Books, 1999), 59–94.

24. Howard Taubman, "Not What it Seems," *New York Times,* November 5,
1961, II 1.

25. Ibid.

26. See, for example, "Who's Afraid of Little Annie Fanny?" *Ramparts* February 5, 1967, 27–30.

27. Edward Albee *The Zoo Story* in *Two Plays by Edward Albee* (New York:
Signet Books, 1961), 25.

28. "The Third Sex On Stage," *Show Business Illustrated,* April 1962, 78.

29. Richard Schechner, "Comment," *Tulane Drama Review,* 7, no. 3 (1963): 9.

30. Qtd. in Gussow, 220. Even some gay critics suspected Albee of disguising
homosexual characters or intent beneath heterosexual characters. Georges-
Michael Sarotte claimed that *Virginia Woolf* is "a homosexual play from
every point of view, in all its situations and in all its symbols." It is only a
"heterosexual play" in its most outward manifestations. He adds that, as is
true of all homosexual writers, Albee's "obsessions emerge, despite all he
can do, to color and contravene his best intentions." Indeed, he finds all of
Albee's work to be gay. (Sarotte, 137–49).

31. Donald M. Kaplan, "Homosexuality and American Theatre: A Psychoanalytic Comment," *Tulane Drama Review,* 9, no. 3 (1965): 25–55.

32. Howard Taubman, "Modern Primer," *New York Times,* April 28, 1963,
II 1.

33. Martin Gottfried, "Theatre," *Women's Wear Daily,* November 12, 1965.
Other references are in the text.

34. Qtd. in Allan Pierce, "Homophobia and the Critics," *Christopher Street,*
June 1978, 41.

35. Stanley Kauffmann, "Homosexuality," in *Persons of the Drama* (New York: Harper and Row, 1976), 291–2. Further references are in the text.

36. The play was a sensation abroad, as well. James Baldwin, living in Istanbul, decided to direct a production of it himself. *Variety's* coverage of the event reflected the confusion of criticism, disdain, virtue, and deep uncertainty with which plays dealing (in this case, indirectly) with homosexuality were greeted in the middle and late 1960s. The first performance, the paper reported, took place for a gala audience of "society people and left-wing intellectuals" who gave the play and director a long ovation. The story was presented with a typically snappy *Variety* headline: "Shock-Sensation in Istanbul as Homo Play on Prisons Directed by Baldwin" (Refif Erduran, *Variety,* January 14, 1970, 91).

37. Jerry Tallmer, "Life Sentence," *New York Post,* March 15, 1967.

38. Charles Dyer, *Staircase* (New York: Grove Press, 1969), 34, 32.

39. Walter Kerr, "Honest, Human," *New York Times,* January 21, 1968, D3; Otis L. Guernsey, *Curtain Times: The New York Theatre 1965–1987* (New York: Applause, 1987), 98.

40. Donald Vining, *A Gay Diary Volume Four, 1967–1975* (New York: Pepys Press, 1983), 28. When Vining saw Joe Orton's *Loot* two months later, he wrote, "There is quite a play in the author's life but I wouldn't try to write it because the public already thinks of the homosexual life as more violent and wretched than it is" (33).

41. Jerry Tallmer, *New York Post,* September 27, 1968.

42. Martin Gottfried, "Theatre: 'The Boys in the Band'," *Women's Wear Daily,* April 16, 1968, 44.

43. Clive Barnes, "Theatre: 'Boys in the Band' Opens Off Broadway," *New York Times,* April 15, 1968, 48; Guernsey, 98.

44. George Oppenheimer, "'Boys in the Band,'" *Newsday,* April 15, 1968. When the play opened in London in 1969, Ronald Bryden in the *Observer* felt as free as the New York critics to announce his prejudices: "At Wyndham's, The Boys in the Band fails to disguise behind whizbang Broadway production a basically cheap and cheapening tourist bus-tour of Fagsville, New York, with the natives wearing their traditional homosexual costumes, dancing their tribal dances and generally camping it up for the visitors. The final message is that everybody is queer, but the ones who admit it are nicer." (*Observer,* February 6, 1969)

45. Michael Smith, *Village Voice* April 25, 1968.

46. Vining, *Gay Diary* Vol. 4, 36–37.

47. Sam Zolotow, "'Boys in the Band' Increases Prices," *New York Times,* November 6, 1968, 34; Sam Zolotow, "Cohen to Present Plays in London with Eye on U.S.," *New York Times,* April 25, 1968, 52.

48. Vining, *Gay Diary* Vol. 4, 34–5.

49. Williams: I think that everybody has some elements of homosexuality in him, even the most heterosexual of us.

Frost: That no one is all man or all woman, you mean?
Williams: Oh, in my experience, no. I don't want to be involved in some sort of a scandal, but I've covered the waterfront (Qtd. in Devlin, 146).

50. Qtd. in Devlin, 189.
51. David Lobdell to Patricia Hepplewhite, December 6, 1970. Rare Book and Manuscript Library, Columbia University.
52. Martin Bookspan, review of *Nightride*, New York: Channel 11 WPIX, December 9, 1971.
53. Lee Barton, "Why Do Homosexual Playwrights Hide Their Homosexuality?" *New York Times*, January 23, 1972, II, 1.
54. Arthur Bell, "'I've Never Faked It,'" *Village Voice*, February 24, 1972, 60.
55. Qtd. in Devlin, 129, 132.
56. Qtd. in Windham, 306–7.
57. Tennessee Williams to David Lobdell, September 15, 1968. Tennessee Williams Papers, Rare Books and Manuscript Library, Columbia University.
58. de Jongh, 70.
59. Tennessee Williams, *Small Craft Warnings* in *The Theatre of Tennessee Williams* Vol 5 (New York: New Directions, 1976), 225. Subsequent references are in the text.
60. Clum, *Acting Gay*, 162.
61. Ibid., 163.
62. Ibid., 164.
63. Spoto, 80.
64. Richard Watts, Jr., *New York Post*, March 3, 1972; Harold Hobson, *Sunday Times*, February 4, 1973; Ted Kalem, "Clinging to a Spar," *Time*, April 17, 1972; Tom McMorrow, "Author Rakes Stage," *New York Daily News*, June 8 1972.
65. Vining, *Diary*, Vol. 4, 25.
66. Edmund White, "Fantasia on the Seventies," *Christopher Street*, September 1977, 19.

CHAPTER 6

1. Qtd. in Devlin, 239.
2. Spoto, 337.
3. David Lobdell letter to Patricia Hepplewhite, October 29, 1970. The observant Lobdell also noted the demons that still drove Williams. "That man can't seem to stay still," he wrote his sister on October 12. "It's almost

as if he were being driven by forces that he can't identify but which he feels he must remain far in front of."

4. *This Is* takes place at the Grande Hotel Splendide in a small European country being besieged by rebels, and includes a character named Count Rechy, who is accused by the Countess of being a repressed, closeted sodomite (the name may have been a joking reference to John Rechy, the young writer whose first book about the gay sexual underground, *City of Night*, had become a cult best-seller and was himself anything but repressed and closeted. Williams did not know Rechy, but his following of the gay press, if not his habitual wide reading, had certainly made him acquainted with the Rechy's exploits).

5. Spoto, 342; Gene Persson, telephone interview. September 16, 2002.

6. Edmund White, telephone interview. October 21, 2002.

7. "Tennessee Williams, On Sexual Identities in His Plays," in *Gay Roots: Twenty Years of Gay Sunshine*, ed. Winston Leyland (San Francisco: Gay Sunshine Press, 1991), 326.

8. George Whitmore, "Tennessee Williams," in *Gay Sunshine Interviews*, Vol. 1 (San Francisco: Gay Sunshine Press, 1984), 313. Subsequent references are in the text. *Memoirs*, 51.

9. Certainly, Williams didn't advance his cause among many gay people by holding forth on transvestites in his *Gay Sunshine* interview: "I don't understand most transvestites. I think the great preponderance of them damage the gay liberation movement by travesty, by making a travesty of homosexuality, one that doesn't fit homosexuality at all and gives it a very bad public image. We are *not* trying to imitate women. We are trying simply to be comfortably assimilated by our society" (312). Not too many months before this, the director and a cast of four actors were arrested in Atlanta after a performance of David Gaard's gay play, *And Puppy Dog Tails*. The director, who was also a coproducer, was charged with allowing an indecent show and the actors with public indecency. In the light of such incidents, which were not infrequent in the 1970s, many gay men and women, not necessarily activists, might have wondered just who Williams's "we" were, who simply wanted to be "comfortably assimilated" (*Variety*, February 12, 1975).

10. Harold M. Schmeck, Jr., "Psychiatrists Approve Change on Homosexuals." *New York Times*, April 9, 1974, 12.

11. Al Carmines, "'Politics is Not Art.'" *New York Times*, July 29, 1973 II, 12.

12. Martin Duberman, "The Gay Life: Cartoon vs. Reality?" *New York Times*, July 22, 1973 II 1, 4.

13. "Furor Over 'The Faggot'," *New York Times*, August 12, 1973 II, 8.

14. Ibid.

15. David Lobdell letter to Patricia Hepplewhite, September 29, 1970.

16. Tennessee Williams, *Vieux Carré* in *The Theatre of Tennessee Williams* Vol. 8 (New York: New Directions, 1992), 65, 20, 30. Subsequent references are in the text.
17. Spoto, 397.
18. Tennessee Williams, *Something Cloudy, Something Clear* (New York: New Directions, 1995), 59. Subsequent references are in the text.
19. Williams used a different version of this same Rilke quote (or perhaps he just wrote it from memory) to end *Now and At the Hour of Our Death*. It's spoken by the Manager, left alone on the stage after Dave has stumbled off to the men's room:

> The Manager
> Who would a character called The Manager be but a mouth-piece of the playwright? Half an hour before curtain-time, he came to my dressing-room, drunk, of course, and handed me a rewrite: *this!*
> (*He holds up a slip of paper.*)
> "Forget the wry comments," he said, "Never mind adding a touch of astringency to the end of this elegy with clowns. Just step through the window onto the forestage and read these concluding lines of the tenth and last of those elegies that the poet Rilke wrote in a storm on a sea-cliff at Schloss-Duino in 1911. The writing is mostly illegible! [This last sentence and the next two words seem to be The Manager's interjection. MP]—Something about ["] the Hero, the youthfully-dead. A Lament, and a Sphinx at the edge of a desert that silently poises for ever the human face on the scale of the stars . . ."
> (*He crumples the slip of paper and drops it: adjusts the white carnation in his button-hole. Removes a pocket-mirror from a vest-pocket and passes it delicately through his thinning hair. Then he turns toward the arch and calls out -*)
> Young man?
> CURTAIN (37)

Selected Bibliography

BOOKS

Bentley, Eric. *What is Theatre?* New York: Limelight Editions, 1984.
———. *The Brecht Memoir.* Evanston: Northwestern University Press, 1989.
Bergler, Edmund. *Homosexuality: Disease or Way of Life?* New York: Hill and Wang, 1957.
———. *One Thousand Homosexuals: Conspiracy of Silence, or Curing and Deglamorizing Homosexuals?* Paterson: Pageant, 1959.
Bérubé, Allan. *Coming Out Under Fire: The History of Gay Men and Women in World War Two.* New York: Plume, 1991.
Boxill, Roger. *Tennessee Williams.* New York: St. Martin's Press, 1988.
Clum, John. *Acting Gay: Male Homosexuality in Modern Drama.* New York: Columbia University Press, 1992.
Clurman, Harold. *The Divine Pastime.* New York: Macmillan, 1974.
Corber, Robert J. *Homosexuality in Cold War America: Resistance and the Crisis of Masculinity.* Durham: Duke University Press, 1997.
Curtin, Kaier. *"We Can Always Call Them Bulgarians."* Boston: Allyson Publications, 1987.
D'Emilio, John. *Sexual Politics, Sexual Communities: The Making of a Homosexual Minority in the United States.* Chicago: University of Chicago Press, 1983.
———. *Making Trouble: Essays on Gay History, Politics and the University.* New York: Routledge, 1992.
Devlin, Albert J., ed. *Conversations with Tennessee Williams.* Jackson: University of Mississippi Press, 1986.

258 ■ *Gentlemen Callers*

Isay, Richard. *Being Homosexual: Gay Men and Their Development.* New York: Avon, 1989.

———. *Becoming Gay: The Journey to Self-Acceptance.* New York: Henry Holt, 1996.

De Jongh, Nicholas. *Not in Front of the Audience.* London: Routledge, 1992.

Katz, Jonathan. *Gay American History: Lesbians and Gay Men in the U.S.A.* New York: Thomas Y. Crowell, 1976.

Kauffmann, Stanley. *Persons of the Drama.* New York: Harper and Row, 1976.

Kubie, Lawrence S. *Practical and Theoretical Aspects of Psychoanalysis.* New York: International Universities Press, 1950.

———. *Neurotic Distortion of the Creative Process.* New York: Noonday Press, 1961.

Leverich, Lyle. *Tom: The Unknown Tennessee Williams.* New York: Crown, 1995.

Lewes, Kenneth. *The Psychoanalytic Theory of Male Homosexuality.* New York: Simon and Schuster, 1988.

Murphy, Brenda. *Tennessee Williams and Elia Kazan.* Cambridge: Cambridge University Press, 1992.

Poland, Albert and Mailman, Bruce, eds. *The Off Off Broadway Book: The Plays, People, Theatre.* Indianapolis: Bobbs-Merrill, 1972.

Roscoe, Will, ed. *Radically Gay: Gay Liberation in the Words of its Founder Harry Hay.* Boston: Beacon Press, 1996.

St. Just, Maria, ed. *Five O'Clock Angel: Letters of Tennessee Williams to Maria St. Just,* New York: Penguin, 1990.

Sarotte, Georges-Michel, *Like a Brother, Like a Lover: Male Homosexuality in the American Novel and Theatre from Herman Melville to James Baldwin.* New York: Anchor Press/Doubleday, 1978.

Savran, David. *Communists, Cowboys and Queers: The Politics of Masculinity in the Works of Arthur Miller and Tennessee Williams.* Minneapolis: University of Minnesota Press, 1992.

Spoto, Donald. *The Kindness of Strangers: The Life of Tennessee Williams.* New York: Ballantine, 1986.

Tischler, Nancy M. *Tennessee Williams: Rebellious Puritan.* New York: Citadel Press, 1961.

Vining, Donald. *A Gay Diary, Vol. 2.* New York: Pepys Press, 1980.

———. *A Gay Diary, 1933–1946.* New York: Hard Candy Books, 1996.

Williams, Edwina. *Remember Me to Tom.* New York: G. P. Putnam's Sons, 1963.

Windham Donald, ed. *Tennessee Williams' Letters to Donald Windham, 1940–1965.* Athens: The University of Georgia Press, 1996.

ARTICLES

Clum, John. "'Something Cloudy, Something Clear': Homophobic Discourse in Tennessee Williams," in *Displacing Homophobia: Gay Male Perspectives in Lit-*

erature and Culture. Edited by Ronald R. Butters, John M. Clum, Michael Moon, Durham: Duke University Press, 1989.

Duberman, Martin Bauml. "Hunting Sex Perverts." *Christopher Street* 5, no. 12 (1982), 43–48.

Holditch, W. Kenneth. "The Last Frontier of Bohemia: Tennessee Williams in New Orleans." *Southern Quarterly* 23 no. 2 (985), 12.

Jones, James, H. "Dr. Yes." *The New Yorker,* August 25–September 1, 1997, 99–113.

Kaplan, Donald M. "Homosexuality and American Theatre: A Comment." *Tulane Drama Review* 9, no. 3 (1965), 25–55.

Lilly, Mark. "*The Glass Menagerie* and *A Streetcar Named Desire.*" In *Lesbian and Gay Writing,* edited by Mark Lily. London: Macmillan, 1990.

Smith, Michael. "The Good Scene: Off Off-Broadway," *Tulane Drama Review* 10, no. 4 (1966), 159–76.

White, Edmund. "Fantasia on the Seventies." *Christopher Street* 5, no. 12, September 1977, 18–20.

Whitmore, George. "Tennessee Williams." *Gay Sunshine Interviews,* Vol. 1 Edited by Winston Leyland. San Francisco: Gay Sunshine Press, 1991, 310–325.

PLAYS

Anderson, Robert. *Tea and Sympathy.* New York: Dramatists Play Service, 1983.

Crowley, Mart. *The Boys in the Band.* New York: Samuel French, 1968.

Goetz, Ruth and Augustus. *The Immoralist.* New York: Samuel French, 1980.

Hoffman, William M., ed. *Gay Plays: The First Collection.* New York: Avon, 1979.

WORKS BY TENNESSEE WILLIAMS

Plays

And Tell Sad Tales of the Deaths of Queens . . . Political Stages: Plays That Shaped a Century. Eds. Emily Mann and David Roessel. New York: Applause, 2002.

Auto-Da-Fé. The Theatre of Tennessee Williams Vol. 6. New York: New Directions, 1981.

Camino Real. The Theatre of Tennessee Williams Vol. 2. New York: New Directions, 1971.

Cat on a Hot Tin Roof. New York: Signet Books, 1955.

Cat on a Hot Tin Roof. New York: Dramatists Play Service, 1958.

The Glass Menagerie. The Theatre of Tennessee Williams Vol. 1. New York: New Directions, 1971.

260 ═ Gentlemen Callers

Lord Byron's Love Letter. The Theatre of Tennessee Williams Vol. 6. New York: New
    Directions, 1981.
Now and at the Hour of Our Death. Special Collections of the UCLA Library.
Orpheus Descending. The Theatre of Tennessee Williams Vol. 3. New York: New Di-
    rections, 1971.
Something Cloudy, Something Clear. New York: New Directions, 1995.
Something Unspoken. The Theatre of Tennessee Williams Vol. 6. New York: 1981.
Suddenly Last Summer. The Theatre of Tennessee Williams Vol. 3. New York: 1971.

Other Works by Williams

The Collected Poems of Tennessee Williams. Edited by David Roessel and Nicholas
    Moschovakis. New York: New Directions, 2002.
Memoirs. New York: Anchor Press/Doubleday, 1972.
The Selected Letters of Tennessee Williams Vol. 1. Edited by Albert J. Devlin and
    Nancy M. Tischler. New York: New Directions, 2000.
"Three Players of a Summer Game," in *Collected Stories.* New York: New Direc-
    tions, 1985.
Untitled poem ["Tiger"]. Tennessee Williams Papers, Rare Book and Manuscript
    Library, Columbia University.

Index

Cronyn, Hume, 12

D

Dave and Jack (*Now and at the Hour of Our Death*), 171–173
de Jongh, Nicholas, 73, 106, 108, 146, 195, 227
desperation, 35–37, 46, 143, 183, 217, 231
discrimination, 52, 86
Drury, Allen, 175
Duberman, Martin, 216
Duvenet, Eloi (*Auto-Da-Fé*), 13, 20–24, 38, 40–41, 146
Duvenet, Madame (*Auto-Da-Fé*), 21–22, 24, 40, 146
Dvosin, Andrew, 212
Dyer, Charles, 183–184

E

Eisenhower, Dwight, 63, 72, 82
"Employment of Homosexuals and Other Sex Perverts in Government," 55
equality, 51, 188
Evans, Oliver, 12–13, 190
evasion, 3, 19, 101–102, 132–133

F

Faggot, The (Carmines), 216–217
federal government, policy on homosexuality, 55–57, 63–64
  Executive Order 10450, 63–64, 72

witch-hunt of homosexuals, 60, 63, 233
see also military, policy on homosexuality; Selective Service Act
Fire Island, 58, 62
flight, 38, 68, 76, 78, 80–82, 132–133, 148, 151, 173, 221
Fortune and Men's Eyes (Herbert), 182–183, 188
freedom, 6, 15, 19, 44, 65–66, 74, 89, 132, 159, 181, 219, 228
Freud, Sigmund, 117–118, 124, 129–130, 132, 152
Friedkin, William, 184
Funke, Lewis, 163

G

Gabrielson, Guy, 54
Garden District (Williams), 122, 139, 141–142, 153–155, 159, 174
Gay Academic Union, 214–216
gay activism, 23, 190, 193, 216
gay audiences, reaction to Williams' plays, 157, 187, 189, 192, 194, 204, 217
gay bars, 12, 25, 62, 71, 85, 87, 134, 136
gay characters, 3, 14, 23–24, 49, 72–73, 161, 164, 174, 182–183, 186, 188–190, 218, 225–226, 233, 235
gay liberation movement, 3, 188–189, 193, 195, 197, 205, 207, 212, 214, 217

LaVergne, TN USA
31 March 2010
177787LV00003B/19/A